Reviews and critical articles covering the entire field of normal anatomy (cytology, histology, cyto- and histochemistry, electron microscopy, macroscopy, experimental morphology and embryology and comparative anatomy) are published in Advances in Anatomy, Embryology and Cell Biology. Papers dealing with anthropology and clinical morphology that aim to encourage cooperation between anatomy and related disciplines will also be accepted. Papers are normally commissioned. Originalpapers and communications may be submitted and will be considered for publication provided they meet the requirements of a review article and thus fit into the scope of "Advances". English language is preferred.

It is a fundamental condition that submitted manuscripts have not been and willnot simultaneously be submitted or published elsewhere. With the acceptance of a manuscript for publication, the publisher acquires full and exclusive copyright for all languages and countries.

Twenty-five copies of each paper are supplied free of charge.

Manuscripts should be addressed to

Co-ordinating Editor

Prof. Dr. H.-W. KORF , Zentrum der Morphologie, Universität Frankfurt, Theodor-Stern Kai 7,
60595 Frankfurt/Main, Germany
e-mail: korf@em.uni-frankfurt.de

Editors

Prof. Dr. F. BECK, Howard Florey Institute, University of Melbourne, Parkville, 3000 Melbourne, Victoria, Australia
e-mail: fb22@le.ac.uk

Prof. Dr. F. CLASCÁ , Department of Anatomy, Histology and Neurobiology
Universidad Autónoma de Madrid, Ave. Arzobispo Morcillo s/n, 28029 Madrid, Spain
e-mail: francisco.clasca@uam.es

Prof. Dr. D.E. HAINES, Ph.D., Department of Anatomy, The University of Mississippi Med. Ctr.,
2500 North State Street, Jackson, MS 39216–4505, USA
e-mail: dhaines@anatomy.umsmed.edu

Prof. Dr. N. HIROKAWA, Department of Cell Biology and Anatomy, University of Tokyo,
Hongo 7–3–1, 113-0033 Tokyo, Japan
e-mail: hirokawa@m.u-tokyo.ac.jp

Dr. Z. KMIEC, Department of Histology and Immunology, Medical University of Gdansk,
Debinki 1, 80-211 Gdansk, Poland
e-mail: zkmiec@amg.gda.pl

Prof. Dr. R. PUTZ, Anatomische Anstalt der Universität München,
Lehrstuhl Anatomie I, Pettenkoferstr. 11, 80336 München, Germany
e-mail: reinhard.putz@med.uni-muenchen.de

Prof. Dr. J.-P. TIMMERMANS, Department of Veterinary Sciences, University of Antwerpen,
Groenenborgerlaan 171, 2020 Antwerpen, Belgium
e-mail: jean-pierre.timmermans@ua.ac.be

213
Advances in Anatomy, Embryology and Cell Biology

Co-ordinating Editor

H.-W. Korf, Frankfurt

Editors

F.F. Beck · F. Clascá · D.E. Haines · N. Hirokawa
Z. Kmiec · R. Putz · J.-P. Timmermans

For further volumes:
http://www.Springer.com/series/102

Emmanouil Skouras,
Stoyan Pavlov,
Habib Bendella,
Doychin N. Angelov

Stimulation of Trigeminal Afferents Improves Motor Recovery After Facial Nerve Injury

Functional, Electrophysiological and Morphological Proofs

With 15 figures

 Springer

Emmanouil Skouras
Department of Trauma, Hand,
  Reconstructive Surgery
Universität zu Köln
Köln
Germany

Habib Bendella
Doychin N. Angelov
Institut I für Anatomie
Universität zu Köln
Köln
Germany

Stoyan Pavlov
Department of Anatomy, Histology and
  Embryology
Medical University Varna
Varna
Bulgaria

ISSN 0301-5556   ISSN 2192-7065 (electronic)
ISBN 978-3-642-33310-1   ISBN 978-3-642-33311-8 (eBook)
DOI 10.1007/978-3-642-33311-8
Springer Heidelberg New York Dordrecht London

Library of Congress Control Number: 2012953091

Printed on acid-free paper

Springer is part of Springer Science+Business Media (www.springer.com)

In Memoriam

In memory of our distinguished colleague and friend Prof. Dr. Jürgen Koebke (Department of Anatomy, University Hospital of Cologne, Germany) who died suddenly on February 23, 2012. His father-like affection to students, sound knowledge in anatomy and brilliant teaching will remain in the hearts of thousands physicians and dentists for ever. Jürgen Koebke's endless enthusiasm for biomechanical and surgical research inspired hundreds of anatomy teachers, abdominal and orthopedic surgeons, and traumatologists worldwide for decades.

In Memoriam

Abstract

Recovery of mimic function after facial nerve transection is poor: the successful regrowth of axotomized motoneurons to their targets is compromised by (1) poor axonal navigation and excessive collateral branching, (2) abnormal exchange of nerve impulses between adjacent regrowing axons, and (3) insufficient synaptic input to facial motoneurons. As a result, axotomized motoneurons get hyperexcitable and unable to discharge. Since improvement of growth cone navigation and reduction of the ephaptic cross talk between axons turn out be very difficult, we concentrated our efforts on the third detrimental component and proposed that an intensification of the trigeminal input to axotomized electrophysiologically silent facial motoneurons might improve specificity of reinnervation. To test our hypothesis we compared behavioral, electrophysiological, and morphological parameters after single reconstructive surgery on the facial nerve (or its buccal branch) with those obtained after identical facial nerve surgery but combined with direct or indirect stimulation of the ipsilateral infraorbital (ION) nerve. We found that in all cases, trigeminal stimulation was beneficial for the outcome by improving the quality of target reinnervation and recovery of vibrissal motor performance.

Acknowledgments

This work has been supported by the Jean Uhrmacher-Foundation, the Imhoff-Foundation, and the Köln Fortune Programm. Special thanks to our colleagues and friends Prof. Dr. Athanasia Alvanou, Prof. Dr. Sarah Dunlop, Dr. Maria Grosheva, Dr. Marcin Ceynowa, Prof. Dr. Orlando Guntinas-Lichius, Dr. Peter Igelmund, Privatdozent Dr. Andrey Irintchev, Prof. Dr. Katerina Kaidoglou, Dr. Daniel Merkel, Dr. Anastas Popratiloff, Prof. Dr. Nektarios Sinis, and Prof. Dr. Michael Streppel.

Contents

List of Abbreviations

AchE	Acetylcholine esterase
BBA	Buccal–buccal anastomosis, that is, transection and end-to-end suture of the buccal branch of the facial nerve (enables regrowth of the buccal branch of the facial nerve)
BSA	Bovine serum albumin
CLSM	Confocal laser scanning microscopy
CMAP	Compound muscle action potential
Contra-ION-ex	Excision of the contralateral infraorbital nerve
DiI	$1,1'$-Dioctadecyl-3,3,3′,3′-tetramethylindocarbocyanine perchlorate (red fluorescent dye)
DMSO	Dimethyl sulfoxide
DPO	Days postoperation
EE	Enriched environment
ES	Electrical stimulation
FB	Fast Blue (blue fluorescent dye)
FFA	Facial–facial anastomosis, that is, transection and end-to-end suture of the facial nerve (enables regrowth of the facial nerve)
FG	FluoroGold (yellow fluorescent dye)
HRP	Horseradish peroxidase (retrograde neuronal label)
ION	Infraorbital nerve
ION-ex	Excision of the infraorbital nerve (causing degeneration of ION)
ION-S	Transection and suture of the infraorbital nerve (enables regrowth of ION)
Ipsi-ION-ex	Excision of the ipsilateral infraorbital nerve
LLS	Levator labii superioris muscle
MS	Mechanical stimulation
NSE	Neuron-specific enolase
NSS	Normal sheep serum
SS	Sham stimulation
TBS	Tris-buffered saline
VS	Vibrissal stimulation
WFM	Widefield microscopy

Chapter 1
Introduction

The facial nerve is the most frequently affected nerve in head and neck trauma. Apart from traffic accident injuries (brainstem hemorrhage, temporal bone fractures, or lacerations of the face), most of the facial nerve lesions are postoperative (removal of cerebellopontine angle tumors, acoustic neuroma surgery, or parotid resections because of malignancy). Despite the use of all available microsurgical techniques for repair of transected nerves, the recovery of facial tone, voluntary movement, and emotional expression of the face remains poor (Anonsen et al. 1986; Ferreira et al. 1994; Vaughan and Richardson 1993).

The inevitable occurrence of a "post-surgery paralytic syndrome" including abnormally associated movements and altered blink reflexes (Kimura et al. 1975; Bento and Miniti 1993) has been attributed to (1) "misdirected" reinnervation of the facial muscles (Montserrat and Benito 1988; Sumner 1990), (2) trans-axonal exchange of abnormally intensive nerve impulses between axons from adjacent fascicles (Sadjadpour 1975), and (3) alterations in synaptic input to facial motoneurons (Bratzlavsky and van der Eecken 1977; Graeber et al. 1993; Moran and Neely 1996).

A key issue in these explanations, which do not exclude each other, is the abnormal activity pattern of the axotomized facial motoneurons. On the one side, the increase in resting potential and existence of still functioning axodendritic synapses (Lux and Schubert 1975; Sumner and Watson 1971) renders them hyperexcitable (Eccles et al. 1958; Ferguson 1978). On the other side, the decreased synthesis of transmitter-related compounds (Lieberman 1971) and reduced axosomatic synaptic input (Blinzinger and Kreutzberg 1968) render the axotomized facial motoneurons less excitable upon afferent stimulation and unable to discharge (Titmus and Faber 1990).

In our work over the last 2 decades we hypothesized that the abnormal activity, that is, the axotomy-caused "silence" of the facial motoneurons, could be improved by alterations in the input from the trigeminal sensory nucleus. This hypothesis gains support from anatomical, electrophysiological, and clinical data showing the involvement of the trigeminal system in generation of facial muscle responses and blink reflexes (Moller and Jannetta 1986; Valls-Sole and Tolosa 1989).

E. Skouras et al., *Stimulation of Trigeminal Afferents Improves Motor Recovery After Facial Nerve Injury*, Advances in Anatomy, Embryology and Cell Biology 213, DOI 10.1007/978-3-642-33311-8_1, © Springer-Verlag Berlin Heidelberg 2013

To test our hypothesis we compared behavioral, electrophysiological, and morphological parameters after single reconstructive surgery on the facial nerve (or its buccal branch) with those obtained after identical facial/buccal nerve surgery but combined with direct or indirect stimulation of the trigeminal (infraorbital, ION) nerve. We found that in all cases, trigeminal stimulation was beneficial for the clinical outcome by improving the quality of target reinnervation and recovery of vibrissal motor performance.

Chapter 2
Materials and Methods

Four major experimental sets have been performed using 413 rats to study the effect(s) of:

1. *Mild* indirect stimulation of the trigeminal afferents (by clipping of the contralateral vibrissal hairs)
2. *Intensive* indirect stimulation of the trigeminal afferents (by excision of the contralateral infraorbital nerve)
3. *Mild* direct stimulation of the vibrissal muscles (by massage)
4. *Intensive* direct stimulation of the vibrissal muscles (by electric current)

All animals were young adult (175–200 g) female Wistar rats (strain HsdCpb: WU, Harlan Winkelmann, Borchen, Germany). They were fed standard laboratory food (Ssniff, Soest, Germany), provided with tap water *ad libitum*, and kept on a 12-h light–dark cycle. Experiments were conducted in accordance with the German law on the protection of animals; procedures were approved by the local animal care committee. We used only female rats because testosterone has been shown to beneficially affect peripheral nerve regeneration (Yu and Yu 1983).

2.1 First Major Set: Mild Indirect Stimulation of the Trigeminal Afferents After Combined Surgery on the Infraorbital and Facial Nerves by Removal (Clipping) of the Contralateral Vibrissal Hairs

2.1.1 Animal Groups and Overview of the Specific Methods Used in the First Experimental Set

Forty-eight rats were randomly divided into four groups (Table 2.1).

Each group consisted of 12 rats that were subjected to unilateral combined injury (transection and suture, i.e., anastomosis) of the right facial (FFA) and infraorbital

E. Skouras et al., *Stimulation of Trigeminal Afferents Improves Motor Recovery After Facial Nerve Injury*, Advances in Anatomy, Embryology and Cell Biology 213,
DOI 10.1007/978-3-642-33311-8_2, © Springer-Verlag Berlin Heidelberg 2013

Table 2.1 Experimental design chart depicting animal grouping and procedures

Group of animals	Video-based motion analysis of vibrissae motor performance	Reinnervation pattern of motor end plates in m. levator labii superioris	Synaptic input to the facial motoneurons	Number of retrogradely labeled trigeminal ganglion cells
1. FFA + ION-S only	12	6	6	6
2. FFA + ION-S + VS	12	6	6	6
3. FFA + ION-S + MS	12	6	6	6
4. FFA + ION-S + VS + MS	12	6	6	6

FFA and ION-S indicate transection and end-to-end suture of the facial and infraorbital branch of the trigeminal nerves, respectively. VS = trimming of contralateral vibrissae to maximize ipsilateral vibrissal use for 4 months; MS = manual stimulation for 4 months; VS/MS indicates 2 months of VS followed by 2 months of MS. All 12 animals in each group underwent video-based motion analysis of the vibrissae motor performance. Thereafter, half of the animals were used to study the reinnervation pattern of m. levator labii superioris. The other six rats in each group were used for retrograde labeling of the facial and trigeminal neurons in the brainstem

(ION-Suture, ION-S) nerves. Rats from group 1 underwent surgery, but received no postoperative therapy (FFA + ION-S-only), whereas rats from groups 2–4 were used to assess the efficacy of three treatment paradigms:

1. Removal of the contralateral vibrissae to ensure a maximal use of the ipsilateral ones (vibrissal stimulation; FFA + ION-S + VS; group 2)
2. Manual stimulation of the ipsilateral vibrissal muscles (FFA + ION-S + MS; group 3)
3. Vibrissal stimulation followed by manual stimulation (FFA + ION-S + VS + MS; group 4)

No intact controls were included in this major experimental set because our aim was not to compare functional and morphological parameters between surgically treated and intact animals: It is well known that after peripheral nerve injury, recovery of function is poor and never reaches levels comparable to those in intact animals.

Four months after combined surgery, all rats were videotaped to determine vibrissal motor performance during explorative whisking using motion analysis system (Table 2.2).

Thereafter, one half of the animals in each group were used to determine the proportion of mono- and polyinnervated motor end plates (Table 2.3) in the ipsilateral levator labii superioris muscle by means of immunocytochemistry for neuronal class III β-tubulin and histochemistry with α-bungarotoxin (see below).

The animals from the second half were used to establish changes in the synaptic input to the facial motoneurons that projected to the whisker pad muscles (identified by retrograde labeling with FB, Table 2.4) and to document eventual changes in the number of those trigeminal ganglion cells sending their dendrites into the infraorbital nerve.

Table 2.2 Motor recovery after combined facial and trigeminal nerve injury and stimulations

Group of animals	Frequency (in Hz)	Angle at maximal protraction (in degrees)	Amplitude (in degrees)	Angular velocity during protraction (in degrees/s)
1. FFA + ION-S only	6.2 ± 1.0	98.0 ± 8.0	11 ± 4.0	138 ± 37
2. FFA + ION-S + VS	6.4 ± 1.5	77.0 ± 11[a]	28 ± 9.0[a]	439 ± 113[a]
3. FFA + ION-S + MS	6.0 ± 1.0	76.0 ± 9.0[a]	30 ± 11[a]	497 ± 163[a]
4. FFA + ION-S + VS + MS	6.3 ± 1.0	74.0 ± 8.0[a]	32 ± 10[a]	528 ± 107[a]

Biometrics of vibrissae motor performance in rats after transection and suture of the right facial and infraorbital nerves (*FFA + ION-S only*), in rats subjected to FFA + ION-S and postoperative clipping of the contralateral vibrissal hairs, that is, vibrissal stimulation (*FFA + ION-S + VS*), in rats subjected to FFA + ION-S and postoperative mechanical stimulation of the vibrissal muscles (*FFA + ION-S + MS*), and in rats subjected to FFA + ION-S and postoperative vibrissal and mechanical stimulations (*FFA + ION-S + VS + MS*). All groups consisted of 12 animals. Shown are group mean values ± SD. Significant differences between group mean values (ANOVA and post hoc Tukey's test, $p < 0.05$)
[a]From FFA + ION-S only
Values adapted from Bendella et al. (2011)

Table 2.3 Quality of target muscle reinnervation

Group of animals	Monoinnervated motor end plates (percent)	Polyinnervated motor end plates (percent)	Non-innervated motor end plates (percent)	Total number of motor end plates examined
1. FFA + ION-S only	35 ± 7.6	58 ± 8.3	7.3 ± 1.4	1,461 ± 241
2. FFA + ION-S + VS	52 ± 7.1[a]	40 ± 3.4[a]	8.1 ± 2.2	1,509 ± 235
3. FFA + ION-S + MS	53 ± 6.1[a]	40 ± 4.1[a]	7.2 ± 3.1	1,719 ± 143
4. FFA + ION-S + VS + MS	56 ± 3.1[a]	33 ± 10[a]	11 ± 3.6	1,482 ± 332

Innervation patterns of the m. levator labii superioris (LLS) motor end plates. Values are means ± SD, $n = 6$ per group
[a]Differences between animals receiving no treatment (group 1) and those receiving VS, MS or VS/MS (ANOVA and post hoc Tukey's test, $p < 0.05$)
All abbreviations are as in Table 2.1

2.1.2 Combined Nerve Surgery (FFA + ION-S)

All surgery was unilateral and performed on the right side under surgical anesthesia (ketamine/xylazine, 100 mg Ketanest®, Parke–Davis/Pfizer, Karlsruhe, Germany, and 5 mg Rompun®, Bayer, Leverkusen, Germany, per kg body weight; i.p.).

Facial–Facial Anastomosis (FFA). The facial nerve was exposed, transected close to its emergence from the stylomastoid foramen, and sutured end to end with two 11–0 atraumatic sutures (Ethicon, Norderstedt, Germany) (Fig. 2.1a).

Transection and Suture of the Infraorbital Nerve (ION-Suture, ION-S). The infraorbital branch (infraorbital nerve, ION) of the ipsilateral trigeminal nerve was exposed at its exit from the infraorbital foramen. All its 4–6 peripheral fascicles were

Table 2.4 Synaptic input to the facial nucleus as estimated by means of widefield fluorescence microscopy (WFM) and confocal laser scanning microscopy (CLSM)

Group of animals	Widefield Fluorescence Microscopy		Confocal Microscopy	
	Synaptophysin containing fraction area in percentage	Number of synapses per motoneuron	Synaptic density (per mm)	Number of synapses per motoneuron	Synapticdensity (per mm)
1. FFA + ION-S only	11.84 ± 1.45	16.48 ± 7.05	128.53 ± 35.04	21.19 ± 6.03	124.80 ± 13.21
2. FFA + ION-S + VS	11.76 ± 1.38	16.15 ± 6.75	126.34 ± 35.63	22.11 ± 5.52	128.08 ± 12.92
3. FFA + ION-S + MS	12.75 ± 1.96	16.52 ± 6.29	133.70 ± 40.96	25.12 ± 4.02	132.61 ± 16.96
4. FFA + ION-S + VS + MS	11.47 ± 1.77	16.39 ± 7.10	128.58 ± 36.23	23.49 ± 5.10	127.64 ± 16.16

Counts of synapse number and density within the borders of selected RoIs. Values are means $\pm$ SD, $n = 6$ per group. There were no differences between rats receiving no intervention and those receiving VS, MS, or VS/MS (ANOVA and post hoc Tukey's test, $p < 0.05$). All abbreviations are as in Table 2.1

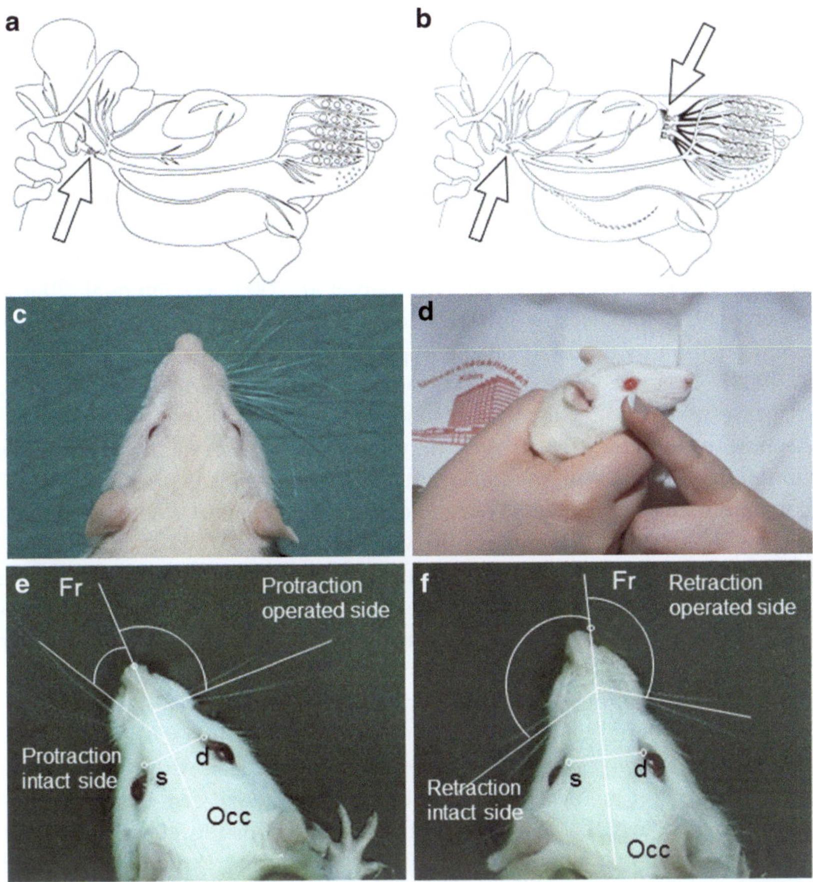

Fig. 2.1 (**a**, **b**) Schematic drawings illustrating transection and end-to-end suture of the infratemporal portion of the facial nerve (**a**, *arrow*) and the infraorbital branch of the trigeminal nerve (**b**, *upper arrow*). Adapted from (Dörfl 1985; Semba and Egger 1986). (**c**, **d**) Postoperative treatments. The vibrissae on the *left* side of the face are trimmed to maximize vibrissal use on the operated, *right*, side (**c**). Manual stimulation of the whisker pad skin and musculature on the operated, *right*, side (**d**).The developed spatial model allows precise measurement of angles, angular velocity, and angular acceleration on the intact (*left*) and operated side (*right*) during protraction (**e**) and retraction (**f**) of the vibrissae. Note the significant change in angle between the sagittal line **Fr-Occ** during protraction and retraction on the intact side. The vibrissae on the operated side remain paretic. Adapted from Tomov et al. (2002)

transected and sutured sequentially one by one (Fig. 2.1b). Theoretically, unilateral lesioning of the infraorbital nerve (ION) could by itself damage ipsilateral vibrissal motor function: The vibrissal system is a sensory–motor loop closed on several levels within sensorimotor structures of the nervous system (Erzurumlu and Killackey 1979; Kleinfeld et al. 1999; Nguyen and Kleinfeld 2005). However, elegant earlier work has shown that profound changes that follow ION injury are not mirrored by the amount of vibrissal muscle motor innervation and function even in newborn rats which are highly sensitive (Veronesi et al. 2006).

2.1.3   Increased Ipsilateral Vibrissal Use (Vibrissal Stimulation, VS) After Combined Surgery in Group 2

On the day following surgery, animals in group 4 (FFA + ION-S + VS) had the contralateral vibrissae clipped (Fig. 2.1c) under light narcosis (Isofluran, O_2 and Niontix: N_2O; Linde AG, Pullach, Germany). Clipping was repeated every third day for 4 months. In this way we hoped to maximize the use of the ipsilateral vibrissae (VS) and therefore to achieve a sensory stimulation of the vibrissal afferent system (Staiger et al. 2000; Hoffman et al. 2003; Machin et al. 2006).

Theoretically, unilateral vibrissae clipping might also trigger an artificially retracted position of the head on the opposite side (Hadlock et al. 2008; Heaton et al. 2008). This phenomenon is observed regularly in all animals that undergo facial nerve transection. Since rats cannot move the vibrissae on the side of the transection (usually the right nerve), they move their heads in a manner allowing maximum exploration with the intact left vibrissae; this leads to a deviation of the head toward the right shoulder. We did not observe increased deviation of the head in animals subjected to trimming of the vibrissae.

Despite clipping of the left whiskers in group 4, our approach should be sharply differentiated from bilateral removal of sensory input: the trigeminal afferents on the side contralateral to FFA are intact and despite clipping 0.5–1.0 mm hair length always remained. Accordingly, animals did not show overt signs of stress, for example, freezing, biting, weight loss, or lack of grooming.

Other possible consequence of depriving animals of normal sensory input on one side of the face is a vigorous whisking on the opposite side. However, we did not observe abnormal use of ipsilateral vibrissae following contralateral clipping, although we watched the animals carefully throughout the entire postoperative period. Rather, animals quickly became accustomed and explored their environment in a manner such that they maximized the likelihood of interactions with objects in their environment while ensuring that such interactions involved only gentle touch (Mitchinson et al. 2007). Thus, the probability to observe a hyperfunction of the contralateral motor facial system, due to abnormally intensive use of the vibrissal hairs, was minimal, and we did not include a separate group of animals with intact facial and trigeminal nerves that had been subjected to enhanced vibrissal stimulation (VS). Since animals could only be videotaped if they had sufficiently long vibrissal hairs on both sides of the face, rats in group 4 were allowed to survive 1 week longer than those in the other groups.

2.1.4   Manual Stimulation of Vibrissal Muscles After Combined Surgery in Groups 3 and 4

In group 3 (FFA + ION-S + MS), MS was initiated on the day following surgery, and in group 4, MS was initiated after 2 months of VS. The right whisker pad (vibrissal hairs, fur, nerves, blood vessels, and vibrissal muscles) were gently and rhythmically stroked by hand for 5 min per day for 5 days a week (Fig. 2.1d) as previously described (Angelov et al. 2007).

2.1.5 Observations on Whisking Behavior

Before each manual stimulation and clipping of the left vibrissae, rats were carefully observed (for about 3–5 min) over the entire course of the experiment.

2.1.6 Analysis of Vibrissae Motor Performance During Exploration

Video-based motion analysis of explorative vibrissal motor performance was performed as described previously in rats (Guntinas-Lichius et al. 2002, 2005b; Tomov et al. 2002) and in mice (Angelov et al. 2003; Guntinas-Lichius et al. 2005a).

The key movements of the vibrissae are protraction (Fig. 2.1e) and retraction (Fig. 2.1f). Since all vibrissal piloerector muscles are innervated by the buccal branch (Dörfl 1985), the whiskers acquire caudal orientation and remain motionless following transection of the facial nerve. Two large vibrissae of the C-row, that is, the third row from dorsal [see Fig. 1 in Arvidsson (1982) and Fig. 2.12b)], on each side of the face were used for biometric analysis, as described previously (Guntinas-Lichius et al. 2001). Under light anesthesia, all other vibrissae were clipped using small fine scissors, and the animals were inserted into a rodent restrainer (Hugo Sachs Electronik—Harvard Apparatus GmbH, 79232 March–Hugstetten, AH 52–0292) for 30 min to pacify them. Employing a digital camcorder (Panasonic NV-DX 110 EG), animals were videotaped for 3–5 min during active exploration. After calibration, video images of whisking behavior were sampled at 50 Hz (50 fields per second); the video camera shutter opened for 4 ms. Images were recorded on AY-DVM 60 EK mini cassettes. Captured video sequences were reviewed and 1.5-s sequence fragments from each animal selected for analysis of whisking biometrics. Thereby, the stable position of animal's head, the frequency of whisking, and the degree of vibrissae protraction were considered as selection criteria.

The tip of the rat's nose and the inner angles of both eyes were defined as reference points. Each vibrissa in the spatial model was represented by 2 points—its base and a point on the shaft 0.5 cm away from the base. Using this model, the following parameters were evaluated: (1) protraction (i.e., the forward movement of the vibrissae) measured by the rostrally opened angle (in degrees) between the midsagittal plane and the hair shaft, (2) the whisking frequency as cycles of protraction and retraction (passive backward movement) per second, (3) the amplitude (the difference between maximal retraction and maximal protraction in degrees), (4) the angular velocity during protraction in degrees per second, and (5) the angular acceleration during protraction in degrees per square second. Measurements were performed by three observers (S.K. Angelova, D. Bösel, D. Felder) who were blinded as to the treatments of the rats.

To clarify the terminology, rodents make a number of different movements with their whiskers, namely, (1) large-amplitude "explorative" sweeps of the vibrissae (in the frequency range of 5–11 Hz), (2) low-amplitude "foveal" or "palpating" whisker movements (at 15–25 Hz; Semba et al. 1980; Berg and Kleinfeld 2003), and (3) denervation-induced tremor that occurs after facial nerve transection (Semba and Komisaruk 1984). All whisker movements are elicited by contractions of the intrinsic and extrinsic vibrissal musculature which are controlled solely by the facial nerve (Dörfl 1985). In this study, we analyzed only the large-amplitude exploratory sweeps. Following facial nerve transection, such exploratory movements are completely abolished. However, with time, and as we show here, there is a gradual progression to varying levels of motor performance which can be readily assessed by video-based motion analysis. The technique is an entirely noninvasive approach to monitor recovery of function after facial nerve repair and avoids the use of invasive electromyography.

2.1.7 Fixation

Fixation. At the end of the designated postoperative survival time, all rats were transcardially perfused with 0.9 % NaCl in distilled water for 60 s followed by a fixation with 4 % paraformaldehyde in 0.1 M phosphate buffer (pH 7.4) for 20 min under deep anesthesia.

2.1.8 Analysis of Target Muscle Reinnervation

Determination of the ratio between mono- and polyinnervated motor end plates was performed as described previously (Guntinas-Lichius et al. 2005b). The levator labii superioris muscles (Fig. 2.2a) were dissected free, cryoprotected in sucrose, and cut longitudinally (30 μm) on a cryostat. Sections were immunostained with a rabbit polyclonal antibody against neuronal class III β-tubulin (Covance, Richmond, CA, USA, No. PRB-435P, 1:1,000) and Cy3-conjugated anti-rabbit IgG (1:400; Sigma). Subsequently, motor end plates were stained with Alexa Fluor 488-conjugated α-bungarotoxin (Molecular Probes, 1:1,000). Quality of end plate reinnervation was evaluated by a simple and straightforward criterion, that is, the number of axonal branches (identified by beta-tubulin staining) that enter or, in some cases, possibly leave the boundaries of individual end plates (identified by acetylcholine receptor staining with alpha-bungarotoxin). Entries by preterminal branches of one axon were counted as single events. According to this criterion, the end plates were identified as "monoinnervated" (one axon; Fig. 2.2c), "polyinnervated" (two or more axons; Fig. 2.2b), or denervated (no visible axonal associated with the receptor staining). The term "polyinnervated end plates" is used to indicate similarity to a morphological abnormality in adult skeletal muscle of mammals observed in pathological conditions such as nerve damage or intoxication that cause axonal branching, which is either

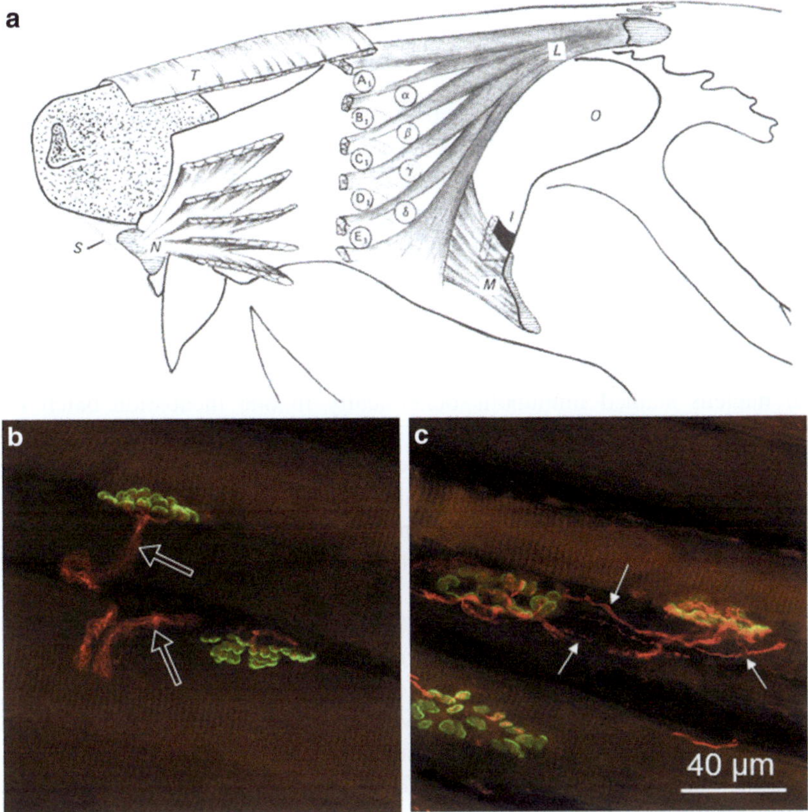

Fig. 2.2 (**a**) Schematic drawing of the extrinsic vibrissae muscles according to Dörfl (1982): α-δ: the four caudal hair follicles, the muscles slings of which "straddle" the five *vibrissae rows* (A–E); T—m. transversus nasi; L—m. levator labii superioris; N—m. nasalis; M—m. maxilolabialis; O—orbit; S—septum intermusculare. (**b, c**) Superimposed stacks of confocal images of end plates in the levator labii superioris muscles of intact and surgically treated rats visualized by staining of the motor end plates with Alexa Fluor 488 α-bungarotoxin (green fluorescence) and immunostaining of the intramuscular axons for neuronal class III β-tubulin (Cy3 red fluorescence). (**c**) and (**b**) show examples of a polyinnervated and a monoinnervated end plate, respectively. Three axonal branches (*arrows* in (**c**)) reach the boundaries of the polyinnervated end plate delineated by the alpha-bungarotoxin staining. In contrast, the monoinnervated end plates in (**b**) are reached by a single axon (*empty arrows* in (**b**)) with several preterminal rami. Adapted from Sinis et al. (2009)

collateral (at nodes of Ranvier) or terminal (from end plate terminals) or both (Rich and Lichtman 1989; Son et al. 1996). Under such conditions many individual end plates are innervated by more than one motoneuron, that is, they are "polyneuronally" innervated. We use the term "polyinnervated end plates" rather than the term "polyneuronally innervated end plates" to indicate that we have not identified the perikaryal origins of supernumerary axons in individual end plates, that is, we did not identify one or more parent motoneurons. Counts of end plates were performed directly under the microscope (objective ×40) in a blind fashion.

2.1.9 Synaptic Input to the Facial Motoneurons

Perikarya in the facial nucleus were visualized by bilateral retrograde labeling with Fast Blue (FB). Under deep ether anesthesia, 1 mg FB (in 100 µl distilled water, 2 % dimethyl sulfoxide) was injected subcutaneously at the midpoint between the two dorsal vibrissae in rows A and B (Arvidsson 1982; Angelov et al. 1993; Angelov et al. 1994; Streppel et al. 1998; Popratiloff et al. 2001) (Fig. 2.3a). Tracer was injected at identical sites in each animal allowing for comparison of synapse numbers on retrogradely labeled perikarya. Sections through the trigeminal ganglia of the same animals were used to examine anatomical restoration of the ION (see below).

Tissue Preparation and Immunocytochemistry. Perfusion-fixed (4 % paraformaldehyde) brainstems were cut (coronal, 30 µm) and every fifth section through the facial nucleus stained immunohistochemically in one incubation batch for all 24 rats. Sections were immunostained with anti-synaptophysin, an established pan-marker for presynaptic terminals, on a shaker at room temperature using (1) 5.0 % (w/v) bovine serum albumin (BSA, Sigma) in TBS for 30 min; (2) 1:4,000 anti-synaptophysin (rabbit polyclonal anti-synaptophysin, Biometra) in TBS plus 0.8 % (w/v) BSA for 2 h; (3) 5.0 % (v/v) normal sheep serum (NSS, Sigma) plus 0.8 % BSA in TBS for 15 min; and (5) anti-rabbit IgG Cy3 conjugate (1:400; Sigma) in TBS plus 0.8 % NSS for 1 h.

Fluorescence Microscopy and Photography. Using a slow-scan CCD camera (Spot RT3 Slider, Diagnostic Instruments, Inc., USA) on a Zeiss Axioplan microscope (Carl Zeiss, Jena, Germany; stabilized powerful UV source: XBO 75 W/HBO100W), images of the facial nucleus from the operated and unoperated side were captured (magnification $\times 10$). We first photographed FB-labeled perikarya ("ultraviolet" filter set 01: excitation BP 365/12, emission LP 397; Carl Zeiss). Synaptic terminals, visualized in red by the CY3 fluorochrome, were then photographed ("rhodamine filter set 15: excitation BP 546/12, emission LP 590; Carl Zeiss). Exposure times were optimized to ensure saturation of only a few pixels. All images were captured under identical conditions. Both picture sets were taken in 14-bpp TIFF format and saved in 16-bpp TIFF format in which every pixel contains 16 bits encoding brightness, ranging from 0 to 65,536, with higher numbers indicating greater brightness (Fig. 2.3b–e).

Image Analysis to Determine Synaptic Density Using Widefield Microsopy. FB-images were used to define "regions of interest" (RoIs) in each picture of the facial nucleus (ImageJ Software v1.38, NIH, Bethesda, Maryland, USA) through the following steps: (1) The dynamic range of the FB-images was maximized using gamma correction ($\gamma = 0.2$). (2) Images were sharpened by subtraction of a blurred copy (Gaussian blurring radius = 75px). (3) Images were automatically thresholded using the Otsu algorithm to produce binary black-and-white images. (4) Motoneurons were included by selecting only FB-labeled areas with a value equal or greater than 500 μm^2. The resulting masks were used to measure the perimeter and area of the selected motoneurons. (5) The perimeters were drawn and expanded in and out by 2px which generated RoIs from the closest perisomatic vicinity of the

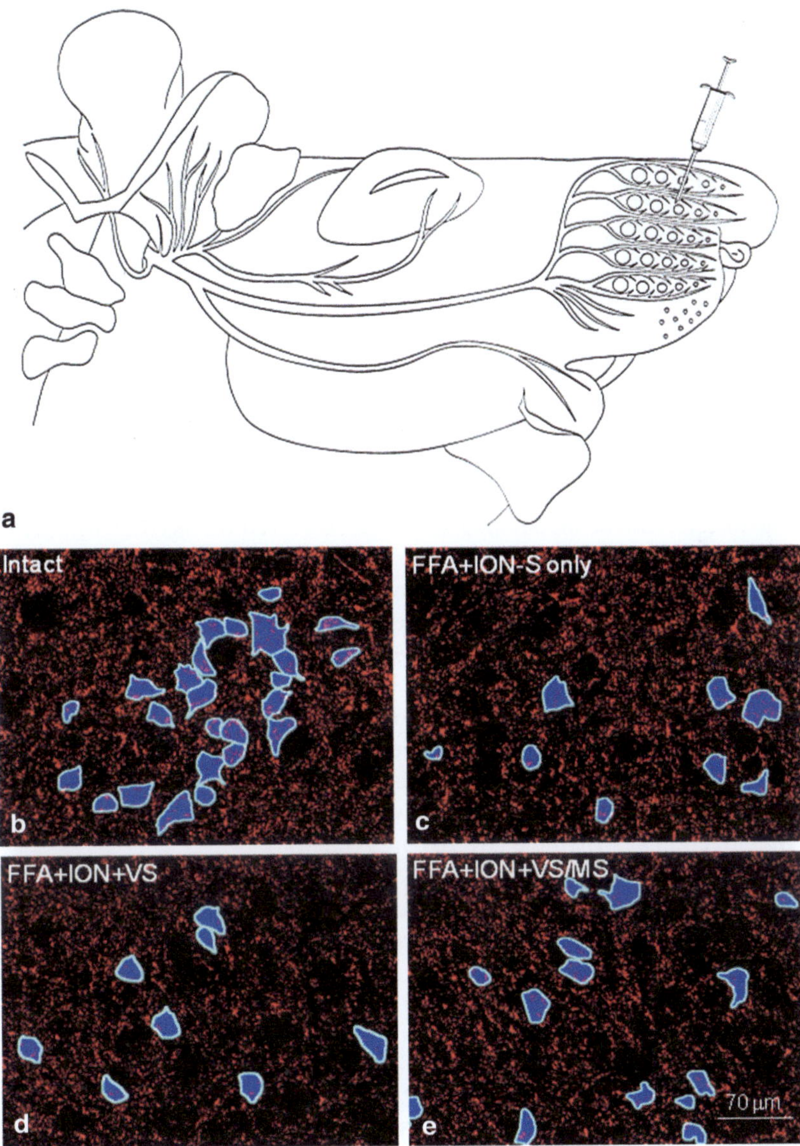

Fig. 2.3 (**a**) Schematic drawing indicating the injection site of the retrograde tracer Fast Blue (syringe in (**a**) into the whisker pad. (**b–e**) Retrogradely labeled motoneuronal perikarya (*blue*) and synaptophysin-CY3 immunostaining (*red*) of axosomatic nerve boutons in the intact facial nucleus (**b**) and 4 months after FFA + ION-S only (**c**), after FFA + ION + VS (**d**), and after FFA + ION + VS/MS (**e**) Scale bar indicates 70 µm

motoneurons with a width of 5px ($\approx$4 µm; Fig. 2.3b–e). All synaptophysin-positive profiles found within each of the predefined perisomatic RoIs of the thresholded images were counted and the "numbers of perisomatic synapses per motoneuron" determined. To correct for size differences, these numbers were weighted with the corresponding perikaryal perimeters to calculate the "linear synaptic density," that is, the number of synapses per mm (Table 2.4).

Image Analysis to Determine the Synaptophysin-Positive Fraction Area in the Lateral Facial Subnucleus. All synaptophysin-positive profiles were processed using a morphological filtering algorithm (top-hat opening by reconstruction), which was similar to the digital background subtraction via granulometric filtering (Prodanov et al. 2006): (1) To define the maximal and minimal radius of the positive profiles before processing, 50 randomly selected images from all experimental sets were subjected to granulometry (data not shown; Grayscale Granulometry for ImageJ, Prodanov D., 2003–2008, http://www.diagnosticarea. com/plugins/gmplugins.html#gran). (2) Next all images were filtered through a minimum filter, the radius of its "wholes" being equal to the minimal profile radius that was measured (6px). This procedure removed all profiles which were smaller than the filter radius. (3) Background was restored using morphological reconstruction (Landini G. GreyscaleReconstruct_v.2.1. for ImageJ, available from http:// www.dentistry.bham.ac.uk/landinig/software. Accessed on 08.07.2010). (4) Background images were subtracted from the originals, which produced images of all positive profiles smaller than the predefined maximal radius (6px). (5) The profile images were binarized by automatic thresholding via the Otsu algorithm. (6) To separate adjacent synaptophysin-positive particles, a "watershed" neighborhood operation was applied to the thresholded images (Fig. 2.4a–d).

To measure the synaptophysin immunoreactivity in the lateral facial subnucleus, all processed profiles were counted by the software, which provided also information on mean area and mean fraction area. This in turn allowed us to calculate the total area of immunopositive profiles as well as the relation of this total area to the area of a given image (particle covered area fraction).

At first glance these stringent features of the method are quite likely to impose substantive biases on the final absolute numbers (i.e., undercounting), mostly arising from the important cell- and synaptic bouton-volume dependence. This undercounting would, however, be the same in all four experimental groups. Furthermore, a quantification of the expression (e.g., intensity of fluorescence) of immune-related antigens by immunocytochemical methods in *absolute values* is considered fraught: The different antigens do not react equally well with the antibodies used (Neefjes and Ploegh 1992). In addition, antibody binding molecules have been observed in large invaginations of the tissue that are not accessible for the incubation medium (Harding et al. 1990). Thus, any quantification of immunopositive cells and cell processes should be considered *relative* and employed only for comparison. Anyway, in analogy to earlier work (Calhoun et al. 1996) we decided to repeat our observations in thin confocal optical sections and thus improve the identification of puncta around the FB-labeled motoneuronal perikarya.

Quantification of Perisomatic Synaptophysin-Positive Puncta. Estimations of perisomatic synaptophysin-positive puncta were performed as described previously (Irintchev et al. 2005). Immunostained sections through the facial nucleus were

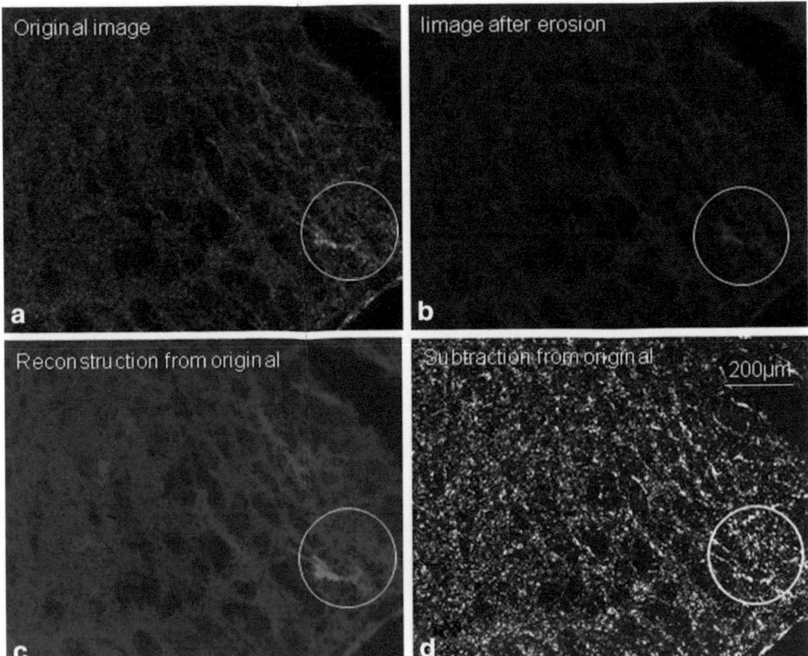

Fig. 2.4 (**a–d**) Steps in image processing of widefield epifluorescence microscopy. (**a**) Original image; (**b**) Erosion of image (**a**) with "minimum filter" with radius 6px. All objects with radius 6px or less are removed and the image is heavily degraded; (**c**) Reconstruction of the eroded image (**b**) from the original (opening by reconstruction). The image is restored to the original state, but all objects with radius 6px or less are missing; (**d**) Subtraction of the reconstructed image from the original (top-hat opening by reconstruction). This operation isolates from the original unprocessed image only the objects with radius 6px or less. Scale bar indicates 200 µm

examined under a fluorescence microscope. Stacks of images of 1-µm thickness were obtained on a TCS SP5 confocal microscope (Leica) using a 40 × oil immersion objective and digital resolution of 1,024 × 1,024 pixels. Four adjacent stacks (frame size, 115 × 115 µm) were obtained consecutively in a rostrocaudal direction so that more motoneurons could be sampled. One image per cell at the level of the largest cell body cross-sectional area was used to count the number of perisomatic puncta (see Fig. 2.5). Motoneurons were easily identified by the retrograde labeling with FB. Areas and perimeters were measured using the Image Tool 2.0 software program (University of Texas, San Antonio, TX). Linear density was calculated as number of perisomatic puncta per unit length. Between 105 and 120 cells were analyzed for synaptic linear density per animal.

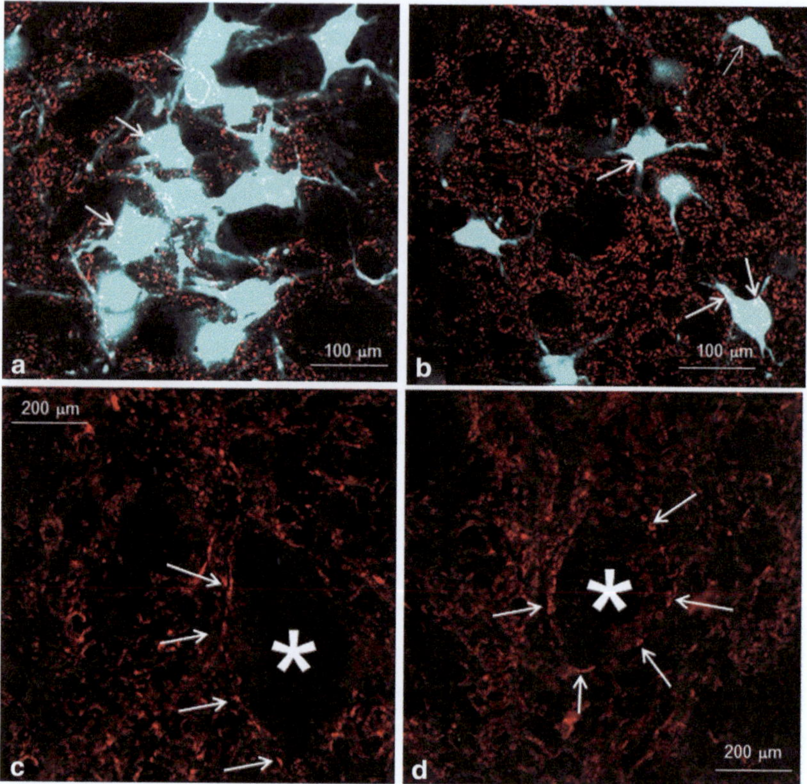

Fig. 2.5 (a–d) Analysis of perisomatic synaptophysin-positive puncta. (**a, b**) Low-power confocal images (1-μm-thick optical slices) show the appearance of synaptophysin-positive puncta (*arrows*) around intact (**a**) and axotomized (**b**) facial motoneurons identified by the retrograde labeling with FB. (**c, d**) High-power confocal images used for counting: intact (**c**) and axotomized (**d**) motoneuronal cell bodies (*asterisks*) 4 months after regeneration following facial and infraorbital nerve transaction and end-to-end suture. Scale bar in (**a, b**) indicates 100 μm, in (**c, d**) 200 μm

2.1.10 Number of Retrogradely Labeled Trigeminal Ganglion Cells

It is well known that trigeminal ganglion neuron numbers vary significantly with age between 3 and 8 months (Lagares et al. 2007), that is, exactly in the period in which we performed our experiments. Nevertheless, we decided to examine the extent of anatomical integrity in the trigeminal ganglion after transection and suture of the infraorbital nerve (ION) and compared the number of retrogradely labeled pseudounipolar trigeminal ganglion cells after injection of FB-solution into the whisker pad (see above). Following perfusion fixation (4 % paraformaldehyde), the

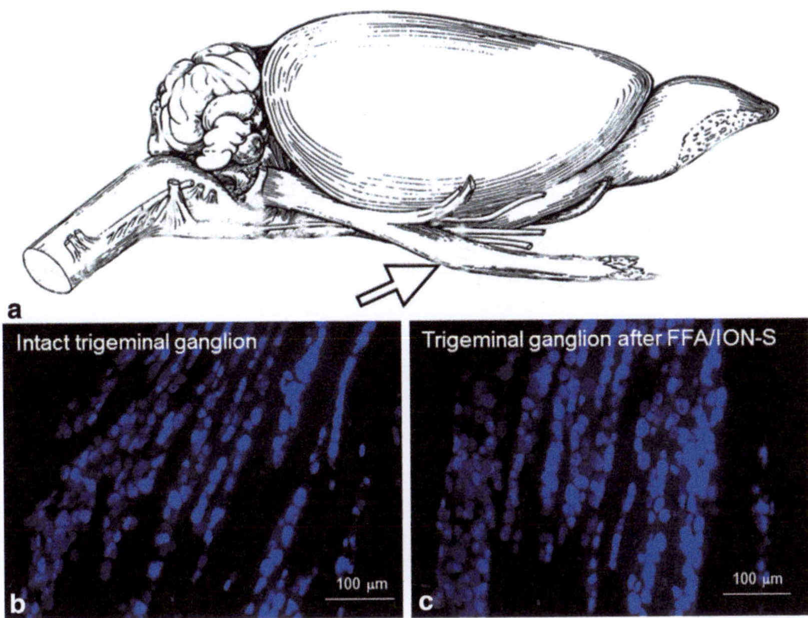

Fig. 2.6 (**a**) Schematic drawing of the trigeminal ganglion (*arrow*), adapted from Greene (1935). (**b, c**) Retrogradely FB-labeled pseudounipolar sensory neurons in an intact trigeminal ganglion (**b**) and in a trigeminal ganglion 4 months after FFA + ION-S only (**c**). No major differences in the amount and distribution pattern of retrogradely labeled cells are evident. 30-μm-thick cryosections. Scale bar indicates 100 μm

trigeminal ganglia were dissected free (Fig. 2.6a), cryoprotected in 30 % sucrose, and cut into 30 thick longitudinal sections which were observed and photographed using the "ultraviolet" filter (set 01, Carl Zeiss).

Since the trigeminal ganglion spans 35–39 sections of 30-μm thickness, every third section was analyzed according to the fractionator selection strategy to determine the total neuron numbers with dendrites projecting into the whisker pad. All retrogradely labeled perikarya with a visible nucleus were counted on the operated and intact sides. Counting was performed blindly with respect to treatment.

2.1.11   Statistics

All data were analyzed using one-way ANOVA with post hoc Tukey's test and significance level of $p < 0.05$. Statistica 6.0 software (StatSoft, Tulsa, OK, USA) was used for analysis.

2.2   Second Major Set: Intensive Indirect Stimulation of the Trigeminal Afferents After Facial Nerve Surgery by Excision of the Contralateral Infraorbital Nerve

2.2.1   Experiments to Determine the Degree of Collateral Axonal Branching by Application of Fluorescent Dyes on the Transected Superior and Inferior Buccolabial Rami of the Buccal Facial Branch

2.2.1.1   Animal Groups and Overview of the Specific Experiments

Seventy rats were divided into seven groups (numbered 1–4 and 2a–4a). Group 1 (10 rats) served as unoperated control.

In groups 1–4, two animals served for electrophysiological measurements and were allowed to survive till 56 days post operation (DPO). The animals of groups 2–4 (each of 14 rats) were used for comparative assessment of axonal regrowth and branching by means of retrograde neuronal labeling. All rats were subjected to identical transection and suture of the right buccal branch of the facial nerve (buccal–buccal anastomosis, BBA). The rats of group 3 underwent BBA plus excision of the ipsilateral (right) ION and those of group 4 BBA plus excision of the contralateral (left) ION. The postoperative survival time was 28 days post surgery and 33–35 days post retrograde labeling. The timecourse was selected on the basis of behavioral observations showing partial restoration of vibrissae rhythmical whisking in some experimental groups during this period.

The animals of groups 2a–4a served to estimate the retraction of axonal branches (pruning or elimination of branches). All rats in groups 2a–4a (each of 6 rats) underwent BBA. The rats of group 3a underwent BBA plus excision of the ipsilateral (right) ION and those of groups 4a BBA plus excision of the contralateral (left) ION (ION-ex). The postoperative survival time was 112 days post surgery and 116–117 days post retrograde labeling.

2.2.1.2   Surgery

Buccal–Buccal Anastomosis (BBA). All operations were carried out under an operating microscope by trained microsurgeons. After intraperitoneal injection of ketamine plus xylazine (100 mg Ketanest® plus 5 mg Rompun® per kg body weight), the buccal branch of the facial nerve was exposed, transected, and immediately sutured with one epineural atraumatic 11–0 suture (Ethicon, Braunschweig, Germany). Since one experimental set of the present study (group C) focused on the accuracy of post-transectional reinnervation by the buccal branch of the facial nerve, we had to eliminate any additional innervation to the whisker pad muscles by the marginal mandibular branch (Semba and Egger 1986). This is the reason why

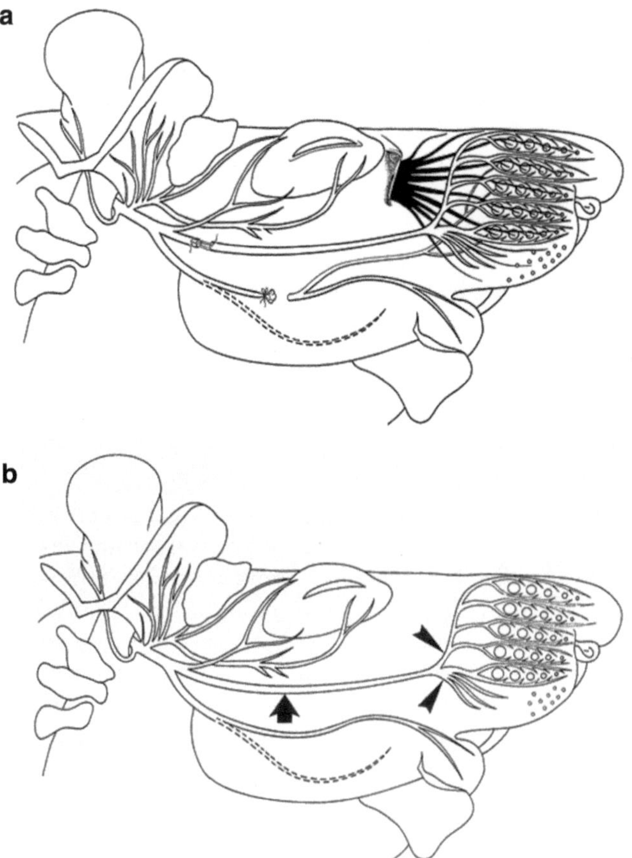

Fig. 2.7 (a) Schematic drawing illustrating the close relationship between the peripheral fascicles of the facial nerve and those of ION (*in black*) and the sites of transection and suture in the buccal branch and of the transection and ligature of the marginal mandibular branch of the facial nerve. The cervical branch of the facial nerve is indicated by a *dotted line*. Adapted from Dörfl (1985) and Semba and Egger (1986). (**b**) Schematic drawing of all fascicles of the infratemporal portion of the rat facial nerve. *Large arrow* indicates the transection and suture site in the buccal branch of the facial nerve. Transection and tracer application sites in the superior and inferior buccolabial nerves are indicated by *arrowheads*

BBA was always accompanied by transection and proximal ligature (to prevent regeneration) of the marginal mandibular branch of the facial nerve (Fig. 2.7a).

Excisions of the infraorbital nerve were performed only in combination with FFA. Under ketamine/xylazine anesthesia, the infraorbital nerve ipsi- or contralateral to the side of FFA was transected at its exit from the infraorbital foramen, and all its peripheral fascicles were removed (resection paradigm; Fig. 2.1a, *upper arrow*).

The aim of this combined facial and trigeminal surgical treatment was to prove whether alterations in the trigeminal input to the axotomized electrophysiologically silent facial motoneurons might improve specificity of reinnervation. The rationale for this approach is derived from existence of direct ipsilateral and "crossed" connections between the trigeminal and facial nucleus (Kimura and Lyon 1972; Erzurumlu and Killackey 1979; Travers and Norgren 1983; Isokawa-Akesson and Komisaruk 1987).

2.2.1.3 Electrophysiological Measurements

In two additional rats of each experimental group, an electroneurography was performed on the right operated facial nerve 7, 28, and 56 DPO. In the control group (1), the measurements were done on day 0, and likewise 7, 28, and 56 days later. The right facial nerve was exposed and a bipolar recording needle electrode was inserted in the middle between the two dorsal vibrissal rows of the ipsilateral whisker pad. The flush-tip monopolar stimulator was placed under the buccal facial nerve directly proximal to the suture site (respectively 4 mm distal to the facial plexus in the two unoperated normal rats). Stimulation was performed for 0.2 ms at a supramaximal intensity of 8.0 mA. The recording analysis time was 50 ms with 10 to 1,000 µV sensitivity and filters of 20 to 3,000 Hz. We measured the amplitude, that is, the peak-to-peak height of the main evoked electromyography waveform, excluding late waves. All measurements were repeated ten times. The data are described as means $\pm$ SD. Group t-tests were used to evaluate the statistical significance of the differences between the experimental groups and the unoperated control group ($p = 0.05$).

2.2.1.4 Estimation of Axonal Regrowth and Branching: Application of Two Crystalline Tracers to Transected Superior and Inferior Buccolabial Nerves

It is well known that the post-transectional misdirection of axons may occur in three ways. *The first way* is that axons are simply misrouted along false endoneural tubes through wrong fascicles toward improper muscles (Esslen 1960; Thomander 1984; Aldskogius and Thomander 1986). *The second way* is that, in contrast with the precise target-directed pathfinding through single motor neurites during embryonic development (Liu and Westerfield 1990), several (but not one single) branches regrow from one transected axon (Shawe 1954; Esslen 1960; Brushart and Mesulam 1980; Ito and Kudo 1994; Baker et al. 1994). *The third way* of misdirection occurs through the intramuscular or terminal sprouting of axons within the target (Son et al. 1996).

In all 70 rats, we performed combined surgery on the facial nerve and ION and studied axonal regrowth peripherally to the transection site, but only within the distal stump of the buccal nerve and its bifurcation. Our aim was, by neglecting terminal intramuscular sprouting (existing, but out of the scope of this chapter), to estimate both regrowth and branching of transected axons (Brown and Hopkins 1981; Jeng and Cogeshall 1984; Duncan and Baker 1987; Brushart 1993). This was quantitatively studied by neuron counts after application of the fluorescent retrograde tracers FluoroGold (FG; Fluorochrome Inc., Englewood, Colorado, 80155, USA) and DiI (1,1′-dioctadecyl-3,3,3′,3′-tetramethylindocarbocyanine perchlorate, Molecular Probes, the Netherlands) to the superior and inferior buccolabial nerves (Fig. 2.7b).

1. In four anesthetized (ketamine/xylazine narcosis) animals of group 1 (unoperated control animals), the superior and inferior buccolabial nerves on the right side of the face were transected and labeled with crystals of DiI and FG, respectively. In the other four animals of group 1, the tracers were interchanged, that is, crystals of DiI were applied to the transected inferior and crystals of FG to the transected superior buccolabial nerve.
2. Identical bilateral labeling was done on day 56 (group 2) or day 112 (group 2a) after unilateral BBA.
3. Identical bilateral labeling was done on day 56 (group 3) or day 112 (group 3a) after unilateral BBA plus excision of the ipsilateral (right) ION.
4. Identical bilateral labeling was done on day 56 (group 4) or day 112 (group 4a) after unilateral BBA plus excision of the contralateral (left) ION.

2.2.1.5 Fixation, Tissue Processing, and Microscopy

Fixation. Five days after the double bilateral labeling (on day 61 or 117 after facial nerve surgery), all rats were transcardially perfused with 0.9 % NaCl in distilled water for 60 s followed by a fixation with 4 % paraformaldehyde in 0.1 M phosphate buffer (pH) 7.4 for 20 min under deep ether anesthesia. After removal of the whole brains, the brainstems were cut coronally in 50-μm sections on a vibratome (FTB-vibracut; Plano, Marburg, Germany).

Fluorescence Microscopy. Vibratome sections were observed through filter set 01 of Carl Zeiss (excitation BP 365/12, emission LP 397), which allows recognition of FG-labeled motoneurons (appearing white). Observations through filter set 15 of Carl Zeiss (excitation BP 546/12, emission LP 590) revealed all motoneurons retrogradely labeled by DiI (appearing red). No fluorescence cross talk was observed between the two tracers used, that is, no DiI-labeled cells were visible through filter set 01 and no FG-labeled motoneurons could be observed using filter set 15. Employing a CCD video camera system (Optronics Engineering model DEI-470, Goleta, CA 93117, USA) combined with the image analyzing software Optimas 6.1 (Optimas Corporation, Bothell, Washington 98011, USA), the image observed with filter set 01 was superimposed on the image taken with filter set 15. The FG/DiI combined pictures of all unlesioned and lesioned facial nuclei in each of 30–33 vibratome sections per animal were saved in a color-coded TIFF format (Fig. 2.8a–f). For manual counting of retrogradely labeled motoneurons, the necessary TIFF file was simply loaded.

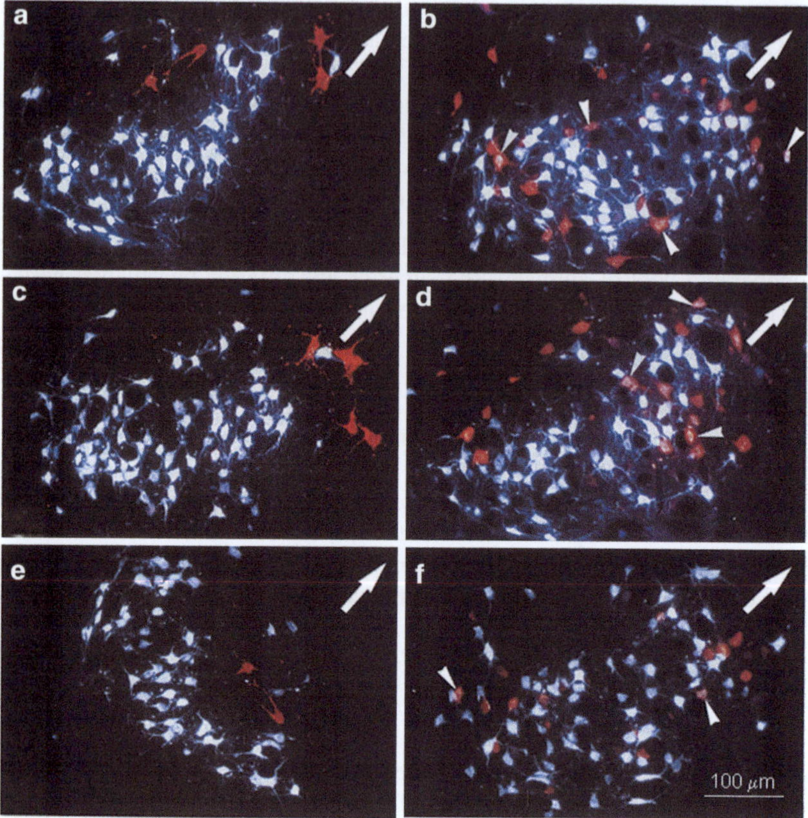

Fig. 2.8 Rat brainstem 28 days after BBA. The dorsomedial portion of the facial nucleus is indicated by an *arrow*. (**a**) Contralateral unlesioned lateral facial subnucleus with preserved myotopic organization of the motoneurons whose axons project into the superior buccolabial nerve (retrogradely labeled in *white* by FluoroGold) and into the inferior buccolabial nerve (labeled in *red* by DiI). Whereas most FG-labeled motoneurons are localized in the ventrolateral portion, those labeled with DiI are in the dorsomedial part of the subnucleus. (**b**) Lesioned lateral facial subnucleus after BBA and application of FG to the superior and DiI to the inferior buccolabial nerve. Note the complete lack of myotopic organization: the FG-labeled (*white*), DiI-labeled (*red*), and DiI + FG-labeled (*arrowheads*) motoneurons are scattered throughout the whole lateral facial subnucleus. (**c, d**) Rat brainstem 28 days after BBA plus excision of the ipsilateral ION. Contralateral unlesioned lateral facial subnucleus with myotopic organization (**c**); Lesioned facial subnucleus 28 days after BBA plus excision of the ipsilateral ION (**d**). (**e, f**) Rat brainstem 28 days after BBA plus excision of the contralateral ION. Contralateral unlesioned lateral facial subnucleus with myotopic distribution of the motoneurons (**e**); lateral facial subnucleus 28 days after BBA plus excision of the contralateral ION (**f**). 50-μm vibratome sections. Scale bar indicates 100 μm

2.2.1.6 Quantitative Estimates

Single postoperative retrograde labeling of facial motoneurons with HRP injected into the whisker pad has shown that the reinnervation of the whisker pad muscles after transection and suture of the main trunk of the facial nerve can be studied in a qualitative and in a quantitative aspect (Thomander 1984; Aldskogius and Thomander 1986; Angelov et al. 1993, 1996). The qualitative aspect is represented by the complete lack of myotopic organization: HRP-labeled motoneurons are scattered throughout the whole facial nucleus. This loss of myotopic organization in the facial nucleus following transection of the peripheral nerve is a direct morphological proof for the occurrence of "misdirected reinnervation" (termed also "misdirected resprouting," "excessive reinnervation," "aberrant reinnervation," "aberrant regeneration," or "misdirected regrowth of axons"). The quantitative aspect of misdirected reinnervation we called hyperinnervation, that is, our counts of HRP-labeled cells, showed that, following facial nerve surgery, there were up to 60 % more motoneurons projecting into the whisker pad muscles than under normal conditions (Angelov et al. 1996). This is why in the present report we also evaluated the reinnervation in a quantitative manner.

Counting. Employing the fractionator principle (Gundersen 1986), all retrogradely labeled motoneurons with visible cell nucleus in the 50-µm-thick sections were counted in every third section through the facial nucleus on the operated and on the unoperated side (Guntinas-Lichius et al. 1993; Guntinas-Lichius and Neiss 1996).

2.2.1.7 Statistics

All data were analyzed as described in Sect. 2.1.11.

2.2.2 *Experiments to Determine the Accuracy of Reinnervation by Means of Intramuscular Injections of Fluorescent Dyes*

Sixty-one rats were used for two types of experiments:

1. Counts after retrograde labeling of facial motoneurons (41 rats)
2. Electrophysiological evaluations (20 rats)

The 41 rats in the anatomical studies were divided into four major groups:

Control Group of 14 Rats. Four rats received 1 % FluoroGold (FG) in the right whisker pad and 1 % Fast Blue (FB) in the left whisker pad. Another four rats received the same injection of FB in the right whisker pad and FG in the left whisker pad. These experiments established whether both tracers had similar efficiency in retrograde neuron labeling. Six other rats received 1 % FG in the right whisker pad, and 56 days later, 1 % FB was injected in the same site. These experiments tested whether the sequential injection of FG and FB in identical muscles can provide a reliable distinction between the FG-, the FB-, and the FG + FB-labeled neurons.

Group FFA. All nine animals received a bilateral intramuscular injection of FG. After 10 days they underwent unilateral FFA. After 56 days, a bilateral postoperative labeling with FB was performed at the site of the earlier FG-injection. The aim of this postoperative labeling was not only to depict the motoneurons projecting into the selected muscles after surgery but to compare their location and number with those of the original innervation pool, which were permanently labeled by the nondegradable tracer FG.

Group FFA Plus Excision of the Ipsilateral ION. All 9 rats underwent identical preoperative labeling with FG. The postoperative labeling with FB was done 56 days after FFA and excision of the ipsilateral infraorbital nerve (ION, ipsi-ION-ex).

Group FFA Plus Excision of the Contralateral (Left) Ion. All nine rats underwent preoperative labeling with FG and postoperative injection with FB 56 days after FFA and excision of the contralateral ION (contra-ION-ex).

2.2.2.1 Preoperative FG-labeling of the Original Motoneuronal Pool

FG was always injected bilaterally. Because we intended to compare the number of retrogradely labeled motoneurons from different animals, great care was taken to ensure identical conditions of injection in all animals. Under diethylether anesthesia, 1 mg FG or 1 mg FB dissolved in 100 µl distilled water containing 2 % dimethyl sulfoxide (DMSO) was injected in the whisker pad muscles at identical sites, at the midpoint between the two dorsal vibrissal rows (Arvidsson 1982). After 10 days, retrograde transport to the facial motoneurons was complete and the rats underwent surgery.

2.2.2.2 Surgery

All surgical operations were performed under an operating microscope. From earlier investigations of the post-transectional misdirection of *axons* projecting into the buccal branch of the facial nerve (Angelov et al. 1999), we learned to ignore the additional innervation of the whisker pad muscles by the marginal mandibular branch (Semba and Egger 1986). In the present study, we focused on the accuracy of *muscle* reinnervation by the buccal branch of the facial nerve, so we had to eliminate additional innervation to the whisker pad muscles. This is why, in all three experimental groups, FFA was always accompanied by transection and proximal ligature (to prevent regeneration) of the marginal mandibular branch of the facial nerve (Fig. 2.7a).

Facial–facial anastomosis (FFA) was performed as already described in Sect. 2.1.2.

Excisions of the ipsilateral or contralateral ION were performed only in combination with FFA. Under ketamine/xylazine anesthesia, the ION was transected at its exit from the infraorbital foramen, and all its peripheral fascicles were removed (resection paradigm).

2.2.2.3   Postoperative Retrograde Labeling of Regenerated Motoneurons

Twenty-eight days post operation, all 27 rats of groups FFA, FFA + ipsi-ION-ex, and FFA + contra-ION-ex received bilateral injections of 1 % FB (1 mg FB in 100 µl distilled water with 2 % DMSO) into the whisker pad musculature, exactly at the same site of the earlier FG-injection.

2.2.2.4   Fixation and Tissue Processing

Ten days after the postoperative bilateral labeling, all rats were transcardially perfused with 0.9 % NaCl in distilled water for 60 sfollowed by a fixation with 4 % paraformaldehyde in 0.1 M phosphate buffer (pH 7.4) for 20 min under deep ether anesthesia. After removal of the whole brains, the brainstems were cut coronally in 50-µm sections on a vibratome.

2.2.2.5   Fluorescence Microscopy

Using standard procedures, FG and FB are both simultaneously visualized with the same UV epifluorescence excitation filter (Zeiss, filter set 01). However, previous experiments have shown that the blue emission of FB obscured the white emission of FG resulting in far too low numbers of FG-labeled neurons. Thus, the quantitative analysis of FG + FB double labeling requires selective custom-made filter sets that exclude most fluorescence cross talk between FG and FB but also reduce sensitivity (Popratiloff et al. 2001). The filter sets used were FG-filter: HQ-Schmalband-filter set (no. F36-050; excitation D 369/40; beam splitter 400DCLP; barrier filter HQ 635/30) and FB-filter: bandpass-filter set (no. F31-000; excitation D 436/10; beam splitter 450 DCLP; barrier filter D470/40) both supplied by AHF Analysentechnik (Tübingen D-72005, Germany).

Employing a CCD video camera system (Optronics Engineering model DEI-470, Goleta, CA 93117, USA) combined with the image analyzing software Optimas 6.5. (Optimas Corporation, Bothell, Washington 98011, USA), separate images of the FG- and FB-retrogradely labeled facial motoneurons were created using these selective special filter sets. The generated masks of FG-labeled cells were superimposed over the FB-image for the unlesioned as well as for the lesioned facial nucleus. With this approach, all cells stained by FG and FB and double labeled by FG + FB could be readily identified and counted (Fig. 2.9c, f, i, l).

2.2.2.6   Quantitative Estimates

Counting. By applying the fractionator principle (Gundersen 1986), all retrogradely labeled motoneurons with visible cell nucleus in the 50-µm-thick sections were counted in every third section through the facial nucleus on the operated and

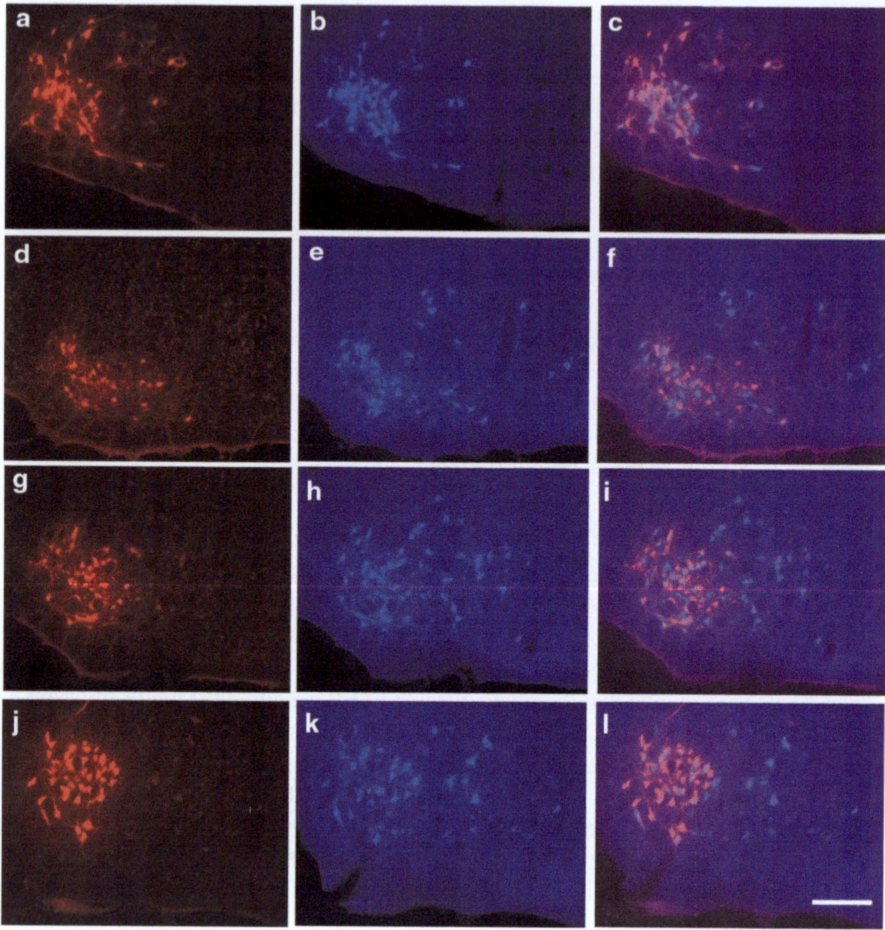

Fig. 2.9 Rat brainstem 28 days after surgery on the buccal branch of the facial nerve. The lateral facial subnucleus, indicated by the preoperative FG-labeling is in the *left* part of each picture. All photographs in the *right* column were produced by double exposure. (**a–b**) Intact facial nucleus with preserved myotopic organization of the motoneurons. Employing the selective filters, we depicted all preoperatively FG-labeled (**a**) and all postoperatively FB-labeled (**b**) motoneurons. (**c**) In the intact facial nucleus, the portion of double-labeled (FG + FB, *pink to bright purple* in color) motoneurons is about 90 %. (**d–f**) Lesioned facial nucleus 28 days after BBA. Whereas all preoperatively FG-labeled motoneurons are localized in the lateral facial subnucleus (**d**), those labeled postoperatively with FB are observed also in the intermediate facial subnucleus (**e**). Our quantitative estimates show that only about 27 % of these FB-labeled motoneurons are double-labeled (**f**) and belong to the original motoneuronal pool of the whisker pad. (**g–i**) Lesioned facial nucleus of a rat 28 days after BBA plus excision of the ipsilateral ION. All preoperatively FG-labeled motoneurons are localized in the lateral facial subnucleus (**g**). The postoperatively FB-labeled motoneurons are found in the lateral and intermediate facial subnuclei (**h**). The double-exposure picture (**i**) is similar to that in (**f**) showing that about 32 % of the FB-labeled cells were also FG-labeled. (**j–l**) Lesioned facial nucleus of rat 28 days after BBA plus excision of the contralateral ION. All preoperatively FG-labeled motoneurons are localized in the lateral facial subnucleus (**j**) and some postoperatively FB-labeled cells are found in the intermediate facial subnucleus (**k**). Our counts show that after this type of combined surgery, the portion of the double-labeled motoneurons (**l**) increased significantly to 41 %. 50-μm-thick vibratome sections. Scale bar indicates 100 μm

unoperated side (Guntinas-Lichius et al. 1993, 2000; Guntinas-Lichius and Neiss 1996). All counts were performed by two observers, who were blind to the surgical procedure used on the rats.

2.2.2.7 Electrophysiological Measurements

Ideally, the electrophysiological test was performed on the same rats that were injected with the retrotracers. This would, of course, provide the most accurate correlation between the two methods. Recently, however, Naumann et al. (2000) provided evidence for possible neurotoxic effects of FG in the medial septal nucleus/diagonal band complex about 6 weeks after retrograde labeling (Naumann et al. 2000). While the previous study focused on a different system and utilized a prolonged "post-labeling" period, we decided to avoid a potential risk of neurotoxicity by conducting the electrophysiological tests on rats that had not been injected with FG. Twenty rats were divided into four groups each consisting of five animals (1) intact rats (control group), (2) rats with unilateral FFA, (3) rats with FFA + ipsi-ION-ex, and (4) rats with FFA + contra-ION-ex. Recordings were performed on day 56 after surgery.

2.2.2.8 Stimulation and Recording

Animals were deeply anesthetized with Nembutal (40 mg/kg body weight) and the body temperature was controlled during the entire recording session. Stimulation and recording were performed with a Neuropack 2 (Nihon Kohden Co., Japan) employing hooked subcutaneous silicon-coated silver wire electrodes (AG-10 T; Science Products GmBH, Germany). The silicone coating at the electrode tip was removed to allow adequate conduction.

Each electrode was inserted into a 20 G needle and hooked over its end. After insertion, the needle was gently retracted, leaving the electrode under the skin. In some animals, this procedure caused bleeding associated with a decline of the recorded signal. In these animals, all further experiments were canceled. The monitor was set to display 20 ms triggered by each stimulus. Recorded signals up to 100 Hz and above 10 kHz were cut off. Stimulation was applied using current mode.

Two stimulation electrodes were placed subcutaneously just in front of the anterior edge of the parotid gland, one above and one below the buccal branch of the facial nerve. The distance between both electrodes was approximately 5 mm. For recordings, a monopolar electrode was placed between the middle vibrissal rows C and D (Arvidsson 1982). Recordings were made with a negative active electrode. On all traces, depolarization of the muscles was indicated by a negative deflection.

To avoid interference from the excitation of other facial muscles, the reference or indifferent recording electrode was placed at a site as close as possible to the recorded ipsilateral vibrissal muscles. However, this procedure could not always be performed perfectly. The innervation domain of the buccal branch, that is, the abundant piloerector muscles plus the levator labii superioris often received a thin communicating branch from the marginal mandibular branch of the facial nerve (Angelov et al. 1999; Semba and Egger 1986). In these cases, we had to attach the indifferent recording electrode proximally to the adjoining point of the communicating branch, which was distal from the whisker pad musculature. Animals were grounded with the aid of a subcutaneous stainless steel needle.

Providing valuable information about the extent of reinnervation and the speed of motor axon conduction, these experiments were designed to evaluate the compound muscle action potential (CMAP) generated by the whisker pad muscles after supramaximal stimulation of the buccal branch of the facial nerve. Once a maximal stimulus had been identified, the stimulation current was increased by 10 %. Usually 7–10 CMAPs were recorded.

2.2.2.9 Quantitative Estimates and Analysis

Two parameters were taken into account: (1) the duration and (2) the amplitude of CMAP. Amplitude was expressed as difference between the maximum peak of CMAP and the baseline (in mV). Duration of CMAP was calculated by the distance between the points where the baseline was crossed by the rising and declining curves of the CMAP. For quantitative and qualitative purposes, the mean duration and amplitude of each experimental group (FFA, FFA + ipsi-ION-ex, FFA + contra-ION-ex) were compared to values obtained in unoperated control animals. The Kolmogorov–Smirnov one-sample test was used to test the normal distribution within the groups. All values are given as means $\pm$ SD (standard deviation). Statistical comparisons of the CMAP measurements among the four groups were performed with ANOVA and followed by a Dunett T3 post hoc test. A P value of less than 0.01 was considered to indicate statistical significance. Unpaired t-test was used to prove whether the mean values for duration and amplitude of CMAPs were significantly different between each experimental group and the control group.

2.3 Third Major Set: Direct Stimulation of the Trigeminal and Facial Nerves After Facial Nerve Surgery by Massage of the Vibrissal Muscles

2.3.1 Animal Groups and Overview of Experiments

One hundred and thirty-eight rats were used with two intact control groups and nine experimental groups (Table 2.5).

Group 1 consisted of 16 intact rats and group 2 of 16 experimental rats which were subjected to unilateral transection and suture of the right facial nerve

Table 2.5 Experimental design flow chart

Group of animals	Video-based motion analysis of vibrissae motor performance	Degree of collateral axonal branching as estimated by triple retrograde labeling	Pattern of reinnervation of motor end plates in m. levator labii superioris
Group 1: Intact animals (16 rats)	16	8	8
Group 2: Animals with right FFA (16 rats)	16	8	8
Group 3: Animals with right FFA + EE (16 rats)	16	8	8
Group 4: Animals with right FFA + right MS (32 rats)	32	8	8
Group 5: Animals with right FFA + EE + right MS (16 rats)	16	8	8
Group 6: Animals with right FFA + left MS (6 rats)	6	–	6
Group 7: Animals with right FFA + handling (6 rats)	6	–	6
Group 8: Animals with right FFA + right ION-ex (6 rats)	6	–	6
Group 9: Animals with right FFA + right ION-ex + MS (6 rats)	6	–	6

Animal grouping and procedures, for example, facial–facial anastomosis (FFA), excision of the ipsilateral infraorbital nerve (ION-ex), dwelling in enriched environment (EE), and mechanical stimulation of the vibrissal muscles (MS). In groups 1–5, the animals that underwent video-based motion analysis were subsequently used for estimation of the degree of collateral axonal branching. In groups 6–9, the animals that were subjected to video-based motion analysis were thereafter used for establishing the pattern of motor end plates reinnervation

(facial–facial anastomosis, FFA) and were allowed to survive for 2 months. In both groups, all rats were used to determine vibrissal motor performance during explorative whisking using video-based motion analysis (Table 2.6). Thereafter, half of the animals were used to establish the degree of collateral axonal branching ipsilaterally by means of retrograde neuronal labeling (Table 2.7). Part of the results obtained for group 1 and 2 have already been published (Guntinas-Lichius et al. 2005b). The remaining eight rats in both groups were used to determine the proportion of mono- and polyinnervated motor end plates (Table 2.8) in the ipsilateral levator labii superioris muscle by means of immunocytochemistry for neuronal class III β-tubulin and histochemistry with alpha-bungarotoxin (see below).

Table 2.6 Recovery of vibrissae function after facial nerve lesion in rats

Group of animals	Frequency (in Hz)	Angle at maximal protraction (in degrees)	Amplitude (in degrees)	Angular velocity during protraction (in degrees/s)
1. Intact	7.0 ± 0.8	62.0 ± 13.2^{a}	57 ± 13^{a}	$1{,}238 \pm 503^{a}$
2. *Right FFA*	*6.3 ± 0.5*	*91 ± 12^{b}*	*19 ± 6^{b}*	*135 ± 54^{b}*
3. Right FFA + EE	6.8 ± 0.9	76 ± 6^{a}	26 ± 5^{b}	490 ± 187^{b}
4a. Right FFA + right MS for 1 min	6.5 ± 0.5 / 6.8 ± 0.9	89 ± 6.2 / 91 ± 10	13 ± 4 / 14 ± 7	175 ± 68 / 159 ± 127
4b. Right FFA + right MS for 2 min	6.6 ± 0.5 / 6.8 ± 0.8	66 ± 15^{a} / 70 ± 11^{a}	51 ± 19^{a} / 36 ± 18^{a}	$1{,}019 \pm 408^{a}$ / 781 ± 329^{a}
4c. Right FFA + right MS for 5 min				
4d. Right FFA + right MS for 10 min				
5. Right FFA + EE + right MS	7.8 ± 2.3	65 ± 16^{a}	55 ± 20^{a}	$1{,}124 \pm 358^{a}$
6. Right FFA + left MS	6.7 ± 1.0	94 ± 9.5^{b}	20 ± 9.5^{b}	368 ± 118^{b}
7. Right FFA + Handling	6.7 ± 0.8	104 ± 10.1^{b}	18 ± 3.4^{b}	316 ± 71^{b}
8. Right FFA + right ION-ex	6.0 ± 0.8	76 ± 10^{b}	22 ± 3.4^{b}	469 ± 400^{b}
9. Right FFA + right ION-ex + MS	6.0 ± 1.2	87 ± 18^{b}	14 ± 5.5^{b}	148 ± 68^{b}

Biometrics of vibrissae motor performance in intact rats (*Intact*), in rats after transection and suture of the right facial nerve only (*right FFA-only*), in rats subjected to FFA and postoperative dwelling in enriched environment (*right FFA + EE*), in rats that were subjected to FFA and postoperative mechanical stimulation of the right vibrissal muscles (*right FFA + right MS*), in rats subjected to combined treatment (*right FFA + EE + right MS*), in rats that were subjected to FFA and postoperative mechanical stimulation of the left vibrissal muscles (*right FFA + left MS*), in rats subjected to FFA and postoperative handling (*right FFA + handling*), and in rats subjected to excision of the ipsilateral infraorbital nerve (ION-ex). Groups 1–3 and 5 consisted of 8 animals, group 4 of 32 rats, and groups 6–9 of 6 animals. Shown are group mean values $\pm$ SD. Significant differences between group mean values (ANOVA and post hoc Tukey's test, $p < 0.05$)
[a]From FFA
[b]From Intact, FFA + MS, and FFA + EE + MS

In groups 3–7, rats underwent unilateral FFA plus subsequent treatments. The animals of group 3 (16 rats) received environmental stimulation for 2 months in an enriched environment (EE). The animals of group 4 (32 rats) received manual stimulation (MS) of the right whisker pad muscles, and the animals of group 5 (16 rats) experienced an enriched environment plus manual stimulation of the right whiskers (EE + MS). Vibrissal motor performance, the degree of axonal branching, and the patterns of motor end plate reinnervation were analyzed.

Groups 6 and 7 consisted of six rats each. Animals in group 6 received FFA on the right side and manual stimulation of the intact contralateral (left) whisker pad muscles. Rats in group 7 received no stimulation of the vibrissal muscles but were

Table 2.7 Projection pattern of facial motoneurons after facial nerve lesion in rats

Group of animals	Neurons projecting only into the zygomatic nerve (*DiI-only*)	Neurons projecting into the zygomatic and buccal nerves (*DiI + FG*)	Neurons projecting into the zygomatic and marginal mandibular nerves (*DiI + FB*)	All DiI-labeled neurons projecting into the zygomatic nerve (*DiI, DiI + FG, DiI + FB*)	Neurons projecting only into the buccal nerve (*FG-only*)	Neurons projecting only into the marginal mandibular nerve (*FB-only*)
Intact[a]	364 ± 47	–	–	364 ± 47	1,441 ± 101	379 ± 94
	100 %	0 %	0 %	100 %		
FFA[a]	*213 ± 53*	*239 ± 52*[b]	*257 ± 56*[b]	*709 ± 178*[b]	*1,908 ± 289*[b]	*1,488 ± 356*[b]
	30 %	*34 %*	*36 %*	*100 %*		
FFA + EE	208 ± 164	140 ± 78[b]	117 ± 76[b]	465 ± 234	2,871 ± 268 [b, c]	2,484 ± 409[b, c]
	45 %	30 %	25 %	100 %		
FFA + MS for 5 min daily	276 ± 219	268 ± 149[b]	211 ± 105[b]	756 ± 251[b]	3,162 ± 342[b, c]	2,614 ± 184[b, c]
	36 %	35 %	29 %	100 %		
FFA + EE + MS	351 ± 178	286 ± 137[b, d]	174 ± 113[b]	810 ± 256[b]	2,790 ± 432[b, c]	1,986 ± 210[b, c, e]
	43 %	35 %	12 %	100 %		

Number of motoneurons with axons in the zygomatic, buccal, or marginal mandibular branches of the facial nerve of intact rats (*Intact*), in rats after transection and suture of the right facial nerve (*FFA-only*), in rats subjected to postoperative dwelling in enriched environment (*FFA + EE*), in rats that received postoperative manual stimulation of the vibrissal hairs (*FFA + MS*), and in rats subjected to combined treatment (*FFA + EE + MS*). The animals were studied 10 days after triple retrograde labeling performed 56 days post surgery. At least eight animals were studied per group. Shown are group mean values ± SD. Significant differences between group mean values (ANOVA and post hoc Tukey's test, $p < 0.05$)

[a]Values adapted from Guntinas-Lichius et al. (2005b)

[b]From Intact

[c]From FFA

[d]From FFA + EE

[e]From FFA + EE and FFA + MS

The percentage values below the absolute numbers in columns 2–5 indicate the portions of motoneurons projecting through the zygomatic nerve with branched (DiI + FG or DiI + FB, column 3 and 4) and unbranched axons (DiI-only, column 2)

Table 2.8 Quality of target muscle reinnervation

Group of animals	Monoinnervated motor end plates (percent)	Polyinnervated motor end plates (percent)	Non-innervated motor end plates (percent)	Total number of motor end plates examined
1. Intact	100 ± 0	0	0	$1{,}543 \pm 132$
2. Right FFA	*45 ± 9.6*	*53 ± 10*	*2.6 ± 1.8*	*1,326 ± 413*
3. Right FFA + EE	50 ± 15	41 ± 15	8.9 ± 5.0^a	$1{,}411 \pm 441$
4 Right FFA + right MS (5 min daily)	$69 \pm 7.9^{a,\,b}$	$22 \pm 5.1^{a,\,b}$	9.6 ± 3.9^a	$1{,}640 \pm 338$
5. Right FFA + EE + right MS	66 ± 11^a	31 ± 10^a	2.7 ± 2.0^c	$1{,}345 \pm 319$
6. Right FFA + left MS	38 ± 7	60 ± 13	2.0 ± 1.6	$1{,}237 \pm 249$
7. Right FFA + handling	39 ± 6	57 ± 12	5.0 ± 2.1	$1{,}402 \pm 235$
8. Right FFA + right ION-ex	51 ± 8.6	43.3 ± 9.4	5.7 ± 2.8	$1{,}495 \pm 435$
9. Right FFA + right ION-ex + MS	41 ± 6.1	50.7 ± 10	8.3 ± 3.6	$1{,}579 \pm 443$

Reinnervation pattern of the levator labii superioris muscle (LLS) motor end plates in intact rats (*Intact*), in rats after transection and suture of the right facial nerve only (*right FFA-only*), in rats subjected to FFA and postoperative dwelling in enriched environment (right *FFA + EE*), in rats subjected to FFA and postoperative manual mechanical stimulation of the right vibrissal muscles (*right FFA + right MS*), in rats subjected to combined treatment (*right FFA + EE + right MS*), in rats subjected to FFA and postoperative mechanical stimulation of the left vibrissal muscles (*right FFA + left MS*), in rats subjected to FFA and postoperative handling (*right FFA + handling*), and in rats subjected to excision of the ipsilateral infraorbital nerve (ION-ex). Motor end plates were classified as monoinnervated, polyinnervated, or non-innervated according to the number of beta-tubulin-immunoreactive axons that crossed the boundaries of the end plate. Groups 1–5 consisted of 8 animals, groups 6–9 of 6 rats. Shown are group mean values $\pm$ SD. Significant differences between group mean values (ANOVA and post hoc Tukey's test, $p < 0.05$)

[a]From FFA
[b]From FFA + EE
[c]From FFA + EE and FFA+MS
Values for intact rats are given as reference values and not included in the analysis

handled by the experimentor in exactly the same way as during MS except that MS was not used ("handling"). Estimation of vibrissal motor performance and patterns of motor end plate reinnervation were estimated 2 months after surgery.

In addition, we took a two-step approach to examine the role of trigeminal afferents (infraorbital nerve) which provide the sensory innervation to the vibrissal muscles (Jacquin et al. 1993; Munger and Renehan 1989; Rice et al. 1993). First, we tested the influence of the trigeminal sensory input on the process of motor recovery by extirpating one of its branches, the infraorbital nerve. The procedure ablates sensory input from the vibrissal muscle pads to facial motoneurons. For this experiment we used two additional groups (Nrr. 8 and 9), each consisting of six rats. In group 8 rats received FFA plus excision of the ipsilateral infraorbital nerve (ION-ex) (FFA + ION-ex). In group 9 rats received FFA plus ION-ex but followed by MS (FFA + ION-ex + MS).

Second, we estimated, using synaptophysin immunohistochemistry, the influence of manual stimulation on the afferent synaptic input to the facial nucleus in three groups of rats (Nrr. 10–12; $n = 6$ in each), namely, intact animals, rats with FFA (FFA-only), and rats with FFA plus MS (FFA + MS). Groups 10–12 are not indicated in Tables 2.5–2.8.

2.3.2   Surgery

Transection and end-to-end suture of the right facial nerve (facial–facial anastomosis, FFA) was performed as described in Sect. 2.1.2.

Excision of the ipsilateral infraorbital nerve was performed only in combination with FFA as described in Sect. 2.2.1.2.

2.3.3   Standard Housing/Enriched Environment

After surgery, all animals were allowed to recover in individual cages for 24 h. Thereafter, rats from groups 2, 4, 6, and 7 were placed in *standard* cages (425 mm × 266 mm × 185 mm; polycarbonate (Tecniplast, Buguggiate, Italy)), each cage with two rats.

All 16 rats in group 3 were placed together in specifically designed cages (three cages sized 610 mm × 435 mm × 215 mm and connected in a row via polycarbonate tunnels, Tecniplast, Buguggiate, Italy) where they experienced group living and an enriched environment consisting of horizontal and inclining platforms and various toys (hanging robes, bridges, tunnels, climbing ladders, balls). Objects and toys were randomly circulated by removing some and adding others during the course of the experiment (cf. van Praag et al. 2000). All 16 rats of group 5 were treated in an identical way, but they received in addition mechanical stimulation of the vibrissal muscles (see below).

2.3.4 Mechanical Stimulation of the Vibrissal Muscles

Mechanical stimulation (both manual as well as environmental) was initiated 1 day after surgery. Rats were daily subjected to gentle rhythmic forward stroking of the right (groups 4 and 5) or left (group 6) vibrissae and whisker pad muscles (Fig. 2.10a,b) 5 days a week. Rats of groups 5 and 6 were manually stimulated for 5 min a day, and rats of group 4 were further distributed into four subgroups (4a, 4b, 4c, and 4d) that were stimulated daily for 1 min, 2 min, 5 min, and 10 min, respectively. The pattern of manual stimulation that we selected mimicked the natural active vibrissal movements during whisking, that is, active protraction and passive retraction (Welker 1964; Wineski 1985). Animals rapidly became accustomed to this procedure within 2–3 days and did not show any signs of stress such a freezing or trying to bite, weight loss, or lack of grooming; rather, animals readily cooperated.

2.3.5 Handling of the Animals

All six rats of group 7 were subjected to daily "handling." Starting from the first day after FFA, animals were carefully taken by an investigator out of the cage and held as if they were to receive MS for 5 min (Fig. 2.10c). Thereafter, rats were put back in the cages.

2.3.6 Analysis of Vibrissae Motor Performance During Exploration

Video-based motion analysis of explorative vibrissal motor performance was performed as described previously in Sect. 2.1.6.

2.3.7 Analysis of the Synaptic Input to the Facial Motoneurons

To assess the synaptic input to the facial nucleus in intact rats and rats subjected to FFA with or without subsequent MS, we quantified levels of synaptophysin according to Calhoun et al. (1996) and Marqueste et al. (2006). Images were obtained on an epifluorescence microscope from sections stained with a highly diluted (1:4,000) anti-synaptophysin antibody. This protocol allowed us to obtain photo images in which, comparable to thin confocal optical sections, numerous puncta within the neuropil and around motoneuronal cell bodies in the facial nucleus were clearly discernible (cf. Fig. 2.3b–e).

Fig. 2.10 Postoperative treatment of the rats. (a) Manual mechanical stimulation of the right vibrissae and whisker pad muscles located ipsilateral to the nerve transection and suture (FFA). (b) Manual mechanical stimulation of the left vibrissae and whisker pad muscles located contralateral to FFA. (c) Handling of the animals

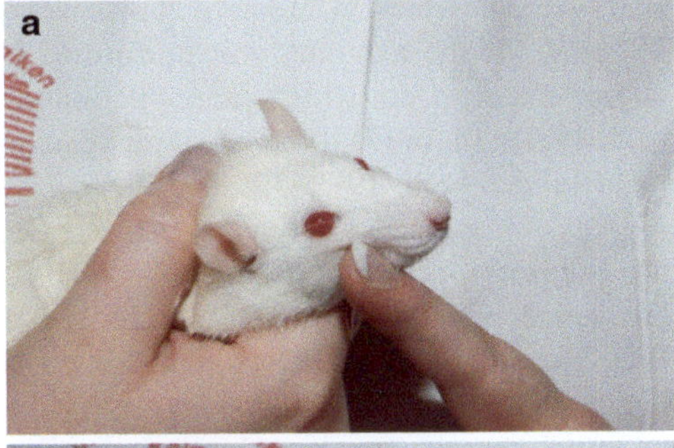
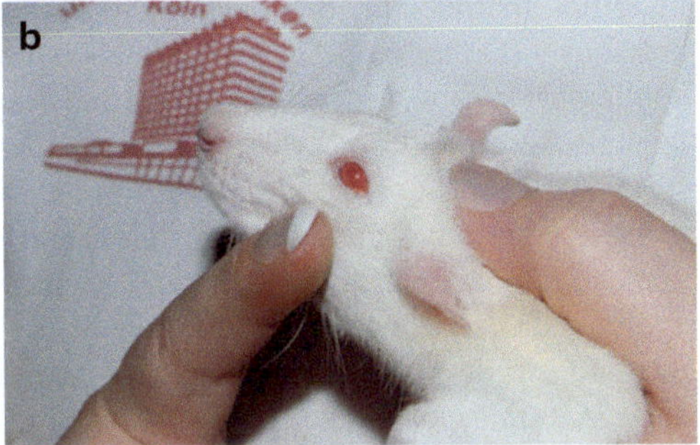
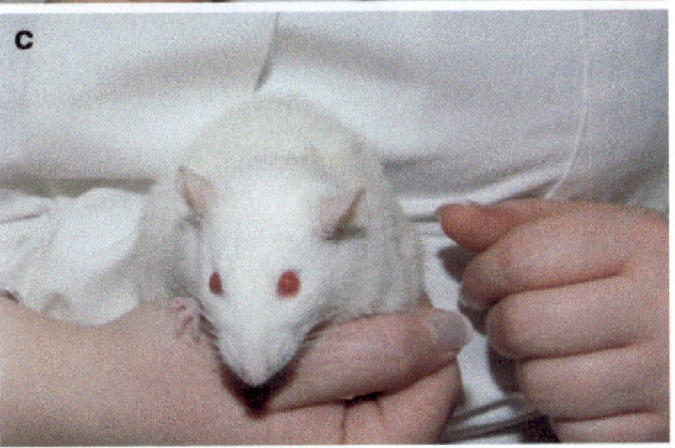

2.3.8   Estimation of Axonal Branching by Triple Retrograde Labeling

Earlier data after immunostaining of 50-μm-thick vibratome sections for neuron-specific enolase (NSE) without retrograde labeling showed that the intact facial nucleus contained $4,066 \pm 508$ NSE-immunoreactive perikarya (Angelov et al. 1994). Similar numbers were found after Nissl staining of paraffin sections ($3,835 \pm 537$ facial neurons; Guntinas-Lichius et al. 1993). After retrograde labeling with horseradish peroxidase, $3,973 \pm 187$ facial motoneurons were found (Angelov et al. 1993).

Subsequent work (Angelov et al. 1994) indicated no significant alterations in these values either at one ($4,152 \pm 166$ facial neurons) or at 8 weeks after FFA ($3,753 \pm 273$ facial neurons). Indeed, it has previously been established that axotomy of the facial nerve causes neuronal cell death only (1) if the rats under study were newborn (Umemiya et al. 1993; Clatterbuck et al. 1994; Rossiter et al. 1996), (2) if the axotomy was performed by removal of the nerve at a length of approximately 1 cm which causes permanent deprivation from the target (Tetzlaff et al. 1988a, b), or (3) if the facial nerve axotomy was performed in mice, but not in rats (Raivich et al. 1998, 1999).

Application of Fluorescence Tracers. One day after videotaping, animals were anesthetized and crystals of DiI (1,1′-dioctadecyl-3,3,3′,3′-tetramethylindocarbocyanine perchlorate, Molecular Probes, Leiden, the Netherlands), FluoroGold (FG; Fluorochrome Inc., Denver, Colorado, USA), and Fast Blue (FB; EMS-Chemie GmbH, Groß-Umstadt, Germany) were applied to the zygomatic, buccal, and marginal mandibular branches of the facial nerve, respectively, as described previously (Dohm et al. 2000). Crystals were left *in situ* for 30 min. Thereafter, the application sites were carefully rinsed and dried and the wound closed. Ten days later, animals were fixed by perfusion with 4 % formaldehyde in 0.1 M phosphate buffer (pH 7.4). Brainstems were sectioned in the frontal plane serially at 50-μm thickness using a vibratome, mounted on slides, and dried at room temperature. The facial nucleus was visible in 32–36 serial sections, and analysis of every third section applying the fractionator principle (see below) allowed estimation of the total number of motoneurons with axons projecting into the ramus zygomaticus, ramus buccalis, and ramus marginalis mandibulae of intact rats (Fig. 2.11; Table 2.7).

Fluorescence Microscopy. Sections were observed with a Zeiss Axioskop 50 epifluorescence microscope (Zeiss, Oberkochen, Germany) using a custom-made bandpass-filter set combination (AHF Analysentechnik, Tübingen, Germany) which restricts the fluorescence cross talk between the tracers *ad maximum.* We used a CCD video camera system (Optronics DEI-470, Goleta, CA, USA) combined with the image analysis software Optimas 6.5. (Optimas Corporation, Bothell, Washington, USA) to create separate color images of retrogradely labeled facial motoneurons through the different filter sets. All images of DiI-labeled motoneurons were used to produce "DiI-masks" using Optimas: Frames were binarized and dilated and the outlines of each DiI-labeled cell depicted. Using the arithmetic options dialogue from the image menu, the masks were superimposed over the FB- or FG-picture. In this way, all cells stained by DiI$_{only}$, FG$_{only}$, FB$_{only}$, as well as all those double stained by DiI + FG or DiI + FB could be readily identified and were manually counted on the computer screen (Dohm et al. 2000). This precise, though time-consuming, procedure allowed us to express the degree (index) of axonal branching in quantitative terms (sum of the percentages given in the third and fourth column in Table 2.7). In rats with an intact facial nerve trunk that had been subjected only to surgery for tracer application, the index of axonal branching was 0 %.

Counting. Employing the fractionator principle (Gundersen 1986), all retrogradely labeled motoneurons with visible cell nucleus in the 50-μm-thick sections were counted in every third section through the facial nucleus on the operated and

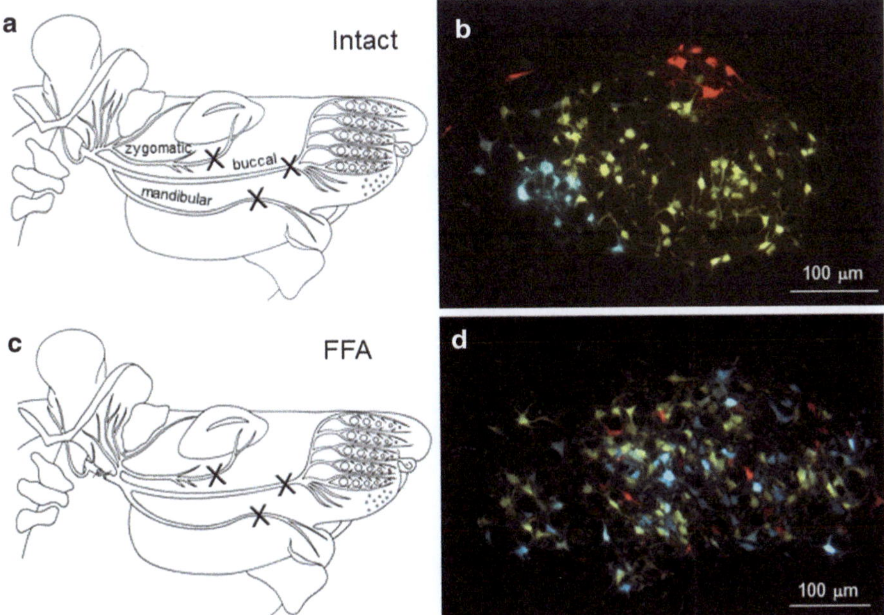

Fig. 2.11 Myotopic organization of the facial nucleus and collateral axonal branching as estimated by the pattern of retrograde labeling. (**a**, **b**) In intact animals, simultaneous application of DiI (*red*), FG (*yellow*), and FB (*blue*) to the zygomatic, buccal, and mandibular nerve branches, respectively, labels distinct subnuclei with no overlap. (**c**, **d**) Two months after transection and suture of the facial nerve, the myotopic organization is lost irrespective whether the animals received ES or not. Adapted from Tomov et al. (2002)

on the intact side. Further details for this sampling technique have been described previously (Neiss et al. 1992; see also note on cell counting on page 98 in Valero-Cabre et al. 2004). Counting was performed blindly with respect to treatment.

2.3.9 Analysis of Target Muscle Reinnervation

Determination of the ratio between mono- and polyinnervated motor end plates was performed as described in Sect. 2.1.6.

2.3.10 Statistical Evaluation

All data were statistically analyzed as described in Sect. 2.1.11.

2.4   Fourth Major Set: Direct Stimulation of the Trigeminal and Facial Nerves After Facial Nerve Surgery by Application of Electric Current to the Vibrissal Muscles

2.4.1   Animal Groups and Overview of Experiments

Ninety-six rats were divided into six groups each comprising 16 animals.

Group 1 consisted of intact rats and groups 2–6 of experimental rats that were subjected to unilateral transection and suture of the right facial nerve (facial–facial anastomosis, FFA; Fig. 2.12a). Animals in group 3 (resection) underwent removal of approximately 1 cm nerve length from the three main branches of the facial nerve (see below). Rats in group 4 (operated, but sham-stimulated, FFA + SS) had electrodes inserted in the denervated vibrissal muscles, but no electric current was applied. In group 5, the vibrissal muscles were subjected to ES (Fig. 2.12c).

Vibrissal motor performance during explorative whisking was analyzed in all rats using video-based motion analysis (Table 2.9). Following the functional analysis at 2 months after operation, half ($n = 8$) of the animals in all groups were used to establish the degree of collateral axonal branching using triple retrograde neuronal labeling (Table 2.10). The remaining rats ($n = 8$) were used to determine the proportion of mono- and polyinnervated motor end plates (Table 2.11) in the largest extrinsic ipsilateral vibrissal muscle, the levator labii superioris (LLS) muscle using immunocytochemistry for neuronal class III β-tubulin, AChE, and histochemistry for alpha-bungarotoxin (see below).

2.4.2   Surgical Procedures

All surgery was unilateral, performed under an operating microscope, and with surgical ketamine/xylazine anesthesia (100 mg Ketanest®, Parke–Davis/Pfizer, Karlsruhe, Germany, and 5 mg Rompun®, Bayer, Leverkusen, Germany, per kg body weight; i.p.).

Transection and suture of the facial nerve (facial–facial anastomosis, FFA) was performed as described in Sect. 2.1.2.

Resection of the Facial Nerve. The main trunk of the facial nerve was unilaterally mobilized at its emergence from the stylomastoid foramen, and pieces of 8–10 mm length were removed from the temporal, zygomatic, buccal, and upper and lower divisions of the marginal mandibular branch. This resection of the facial nerve is a very severe lesion in comparison to crush or transection of the nerve and delivers a permanent separation of the facial motoneurons from their target musculature.

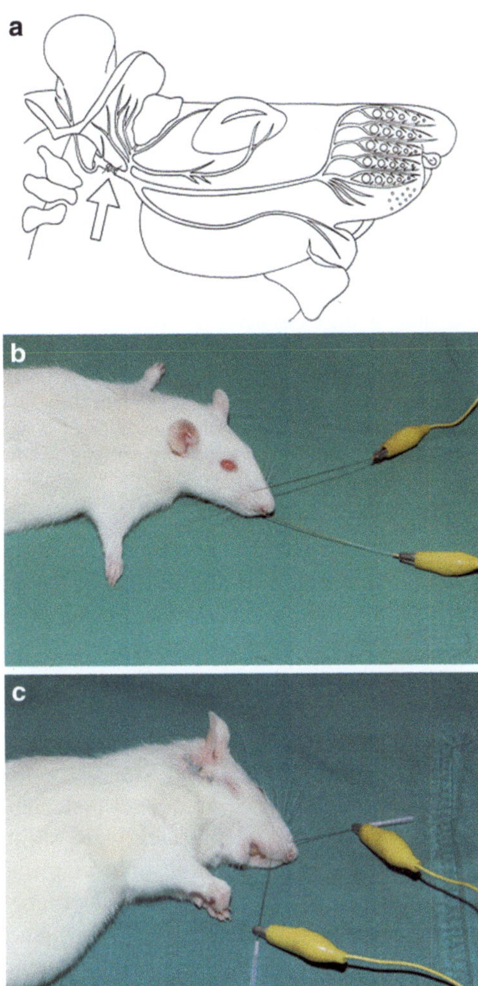

Fig. 2.12 (**a**) Schematic drawing of the infratemporal portion of the rat facial nerve. The site of transection and end-to-end suture of the facial nerve trunk, that is, facial–facial anastomosis (FFA) is indicated by an *arrow*. (**b**) Sham stimulation of rats: Acupuncture needle electrodes were inserted, but no current was applied to the electrodes. (**c**) Postoperative electrical stimulation of the vibrissal muscles. Adapted from Sinis et al. (2009)

2.4.3   Electrical Stimulation

Operated rats were subjected to electrical stimulation (ES) of the vibrissal muscles three times a week (Monday, Wednesday, and Friday) over 2 months starting on the first day after surgery. Electrical stimulation was delivered three times weekly since animals were required to be anesthetized. Under ketamine/xylazine anesthesia, two acupuncture needle electrodes were inserted toward LLS muscle, one along the uppermost vibrissal row A and the other in the lowest row D (Arvidsson 1982). The site of electrode placement close to the nose of the animal was thus somewhat distant (approximately 1 cm) to the location where the majority of LLS motor end plates are found, that is, at the borderline between the cheek and the whisker pad.

Table 2.9 Recovery of vibrissae function after facial nerve injury in rats

Group of animals	Frequency (in Hz)	Angle at maximal protraction (in degrees)	Amplitude (in degrees)	Angular velocity during protraction (in degrees/s)
1. Intact[a]	7.0 ± 0.8	62 ± 13	57 ± 13	1,238 ± 503
2. FFA-only[a]	6.3 ± 0.5	91 ± 12	19 ± 6	135 ± 54
3. Resection	7.0 ± 0.8	102 ± 16	16 ± 5	323 ± 170
4. *FFA + SS*	*5.8 ± 0.7*	*99 ± 11*	*16 ± 2.4*	*323 ± 81*
5. FFA + ES	5.6 ± 1.1	106 ± 33	20 ± 8	211 ± 93
6. FFA + MS[a]	7.8 ± 2.3	65 ± 16	55 ± 20	1,124 ± 358

Biometrics of vibrissae motor performance in intact rats (*Intact*), in rats subjected to transection and suture of the facial nerve (*FFA-only*), in rats that underwent removal of 1 cm length from the main branches of the facial nerve (Resection paradigm), and in rats subjected to FFA plus postoperative sham stimulation (*FFA + SS*), postoperative electrical stimulation (*FFA + ES*), or postoperative mechanical stimulation of the whisker pad (*FFA + MS*). All groups consisted of 16 rats. Shown are group mean values ± SD. No significant differences between the control group 4 (*FFA + SS*) and the group with electrically stimulated rats (ANOVA and post hoc Dunnett's test, $p < 0.05$) were detected
[a]Data adapted from Angelov et al. (2007)

The configuration thus allowed ES of the target muscles, the LLS, and part of the intrinsic vibrissal muscles close to the stimulating electrode, without the risk of direct damage to the motor end plates.

In all electrically stimulated rats ($n = 16$), the threshold voltage required to elicit visible contractions of the whisker pad muscles and movements of the whiskers was initially determined by applying square 0.1 ms pulses at various voltage intensities using an isolated pulse stimulator (Master-8-cp, A.M.P.I., Jerusalem, Israel). The frequency selected (5 Hz) resembled the frequency of normal whisking. The muscles were stimulated for 5 min by applying square 0.1 ms pulses with suprathreshold amplitudes (typically 3.0–5.0 V; Fig. 2.12c). This stimulation was sufficient to depolarize intramuscular axons but not muscle fibers, innervated or denervated, in which action potentials can be elicited only upon much "stronger" stimulation, for example, pulses of 20-V amplitude and 5-ms duration for normal muscles and higher for denervated muscle fibers (Irintchev et al. 1990; Kern et al. 2002). Efficient muscle stimulation, especially that of denervated muscle fibers, requires delivery of high-voltage current pulses of long duration. This stimulation protocol was not approved by the animal welfare committee in Cologne because of the concern that "strong" stimulation might elicit trigeminal pain.

Control sham-stimulated rats were treated identically to rats subjected to ES except that no current was applied to the electrodes (Fig. 2.12b).

2.4.4 Analysis of Vibrissal Motor Performance

Video-based motion analysis of vibrissal motor performance was performed as described previously in Sect. 2.1.6.

Table 2.10 Projection pattern of facial motoneurons after facial nerve lesion in rats

Group of animals	Neurons projecting only into the zygomatic nerve (*DiI-only*)	Neurons projecting into the zygomatic and buccal nerves (*DiI + FG*)	Neurons projecting into zygomatic and marginal mandibular nerves (*DiI + FB*)	All DiI-labeled neurons projecting into the zygomatic nerve (*DiI, DiI + FG, DiI + FB*)	Neurons projecting only into the buccal nerve (*FG-only*)	Neurons projecting only into the marginal mandibular nerve (*FB-only*)
1. Intact[a]	364 ± 47 100 %	– 0 %	– 0 %	364 ± 47 100 %	1,441 ± 101	379 ± 94
2. FFA-only[a]	213 ± 53 30 %	239 ± 52 34 %	257 ± 56 36 %	709 ± 178 100 %	1,908 ± 289	1,488 ± 356
3. Resection	0	0	0	0	0	0
4. *FFA + SS*	*228 ± 165* *43 %*	*159 ± 98* *30 %*	*138 ± 79* *27 %*	*525 ± 342* *100 %*	*2,172 ± 256*	*1,881 ± 409*
5. FFA + ES	321 ± 120 38 %	237 ± 102 28 %	2 76 ± 83 34 %	834 ± 305 100 %	2,254 ± 374	2,156 ± 262
6. FFA + MS[1]	276 ± 219 36 %	268 ± 149 35 %	211 ± 105 29 %	756 ± 251 100 %	3,162 ± 342	2,014 ± 184

Number of motoneurons with axons in the zygomatic, buccal, or marginal mandibular branches of the facial nerve in intact rats (*Intact*), in rats subjected to transection and suture of the facial nerve (*FFA-only*), in rats that underwent removal of 1 cm length from the main branches of the facial nerve (Resection paradigm), and in rats subjected to FFA plus postoperative sham stimulation (*FFA + SS*), postoperative electrical stimulation (*FFA + ES*), or postoperative mechanical stimulation of the whisker pad (*FFA + MS*). The percentage values below the absolute numbers in columns 2–5 indicate the portions of motoneurons projecting through the zygomatic nerve with branched (*DiI + FG or DiI + FB*, column 3 and 4) and unbranched axons (*DiI-only*, column 2). Animals were studied 10 days after triple retrograde labeling. At least eight animals were studied per group. Shown are group mean values ± SD. No significant differences between the control group 4 (*FFA + SS*) and the group with electrically stimulated rats (ANOVA and post hoc Dunnett's test, $p < 0.05$) were detected. Data for groups 1, 2, and 6 are given as reference and are not included in the analysis
[a]Data adapted from Angelov et al. (2007)

Table 2.11 Quality of target muscle reinnervation

Group of animals	Monoinnervated motor end plates (percent)	Polyinnervated motor end plates (percent)	Non-innervated motor end plates (percent)	Total number of motor end plates in LLS muscle
1. Intact rats[a]	100 ± 0	0	0	1,543 ± 132
2. FFA-only[a]	45 ± 9.6	53 ± 10	2.6 ± 1.8	1,326 ± 413
3. Resection	3.1 ± 0.6	1.6 ± 0.4	95 ± 22	416 ± 113
4. *FFA + SS*	*49 ± 7.7*	*49 ± 9.4*	*2.4 ± 0.8*	*1,398 ± 415*
5. FFA + ES	4.4 ± 1.8[b]	5.5 ± 1.4[b]	91 ± 25[b]	346 ± 189[b]
6. FFA + MS[a]	69 ± 7.9	22 ± 5.1	9.6 ± 3.9	1,640 ± 338

Innervation pattern of the levator labii superioris (LLS) muscle motor end plates in intact rats (Intact), in rats subjected to transection and suture of the facial nerve (*FFA-only*), in rats that underwent removal of 1 cm length from the main branches of the facial nerve (Resection paradigm), in rats subjected to FFA plus postoperative sham stimulation (*FFA + SS*), postoperative electrical stimulation (*FFA + ES*), or postoperative mechanical stimulation of the whisker pad (*FFA + MS*). At least eight animals were studied per group. Shown are group mean values ± SD. Data for groups 1–3 and 6 are given as reference and are not included in the analysis
[a]Data adapted from Angelov et al. (2007)
[b]Difference between groups 4 (*FFA + SS*) and 5 (*FFA + ES*) at 2 months after surgery (ANOVA and post hoc Dunnett's test, $p < 0.05$)

2.4.5 *Estimation of Axonal Branching by Triple Retrograde Labeling*

Application of fluorescence tracers, tissue preparation, microscopy, and evaluations were described in Sect. 2.3.7.

2.4.6 *Analysis of Target Muscle Reinnervation*

The ratio between mono- and polyinnervated motor end plates was estimated as described in Sect. 2.1.8.

2.4.7 *Statistical Evaluation*

Data were statistically analyzed as described in Sect. 2.1.11.

Chapter 3
Results

3.1 Mild Indirect Stimulation of the Trigeminal Afferents After Combined Surgery on the Infraorbital and Facial Nerves by Removal of the Contralateral Vibrissal Hairs Improves Vibrissal Function

3.1.1 Observations on Restoration of Vibrissal Whisking

In all groups following FFA + ION-S, the right vibrissae dropped immediately and acquired a caudal orientation. At 10–14 days after FFA + ION-S, the vibrissae of all operated animals (groups 1–4) "rose" to the level of the mouth and acquired a posterior orientation.

In all rats subjected to FFA + ION-S only (group 1), the vibrissal hairs on right side of the face remained motionless with only slight tremor-like movements up to 4 months.

Initial signs of rhythmic whisking in the stimulated rats from groups 2 to 4 were detected at 42–56 days after FFA. No differences in vibrissal whisking could be observed among the rats from the stimulated groups (2–4). This was confirmed by our computerized quantitative estimates.

3.1.2 All Three Interventions (Sensory, Mechanical, and Sensory + Mechanical Stimulation) Improved Vibrissal Function After Combined Facial and Infraorbital Nerve Injury

A critical feature of our video-based assessment is to measure the amplitude of vibrissal bouts during free exploration, thereby avoiding restraint of the animals. Earlier results have shown that, during free exploration, rats rhythmically move

E. Skouras et al., *Stimulation of Trigeminal Afferents Improves Motor Recovery After Facial Nerve Injury*, Advances in Anatomy, Embryology and Cell Biology 213, DOI 10.1007/978-3-642-33311-8_3, © Springer-Verlag Berlin Heidelberg 2013

their heads so as to bring microvibrissae into contact with the surfaces they are exploring. These periodic "microvibrissal placements" are synchronized with macrovibrissal movements not only during free exploration but also during texture discrimination tasks (Quist and Hartmann 2008; Towal and Hartmann 2008). As a result, during our 3–5-min video sessions, natural head movements result in some variability in whisking effort.

However, if the head is immobilized, the finely tuned physiological synchrony of micro- and macrovibrissal whisking is impaired. In an attempt to compensate for the resulting decrease in sensory stimuli, rats react with significantly larger than normal and therefore nonphysiological vibrissal excursions of the order of 90° (Bermejo et al. 1996, 1998; Bermejo and Zeigler 2000; Hadlock et al. 2007).

Four months after combined lesion (FFA + ION-S) without treatment, recovery was poor: Amplitude and angular velocity during whisking were obviously strongly reduced to about 20 % of its value in intact animals (Table 2.2; see also the first bar in Fig. 3.1a for graphical representation of the mean amplitudes).

Four months after FFA + ION-S, all three postoperative stimulations improved function such that the amplitude of vibrissal whisking and angular velocity were significantly higher ($p < 0.05$) than in the surgically treated but non-stimulated animals (Table 2.2; see also the second, third, and fourth bars in Fig. 3.1a for graphical representation). Furthermore, there were no differences between the three treatments. However, despite clear functional improvements, vibrissal whisking amplitude in all stimulated groups was approximately 50 % less (again no statistical analysis was necessary) than values in intact animals (VS: 28° ± 9°; MS: 30° ± 11°; VS followed by MS: 32° ± 10°; Table 2.2).

In normal animals, whisker movements comprise large-amplitude "explorative" sweeps through protraction/retraction cycles (5–11 Hz) and low-amplitude "foveal" or "palpating" movements (15–25 Hz; Semba et al. 1980) with active rostral protraction via muscle contraction (Guntinas-Lichius et al. 2002; Tomov et al. 2002) and active caudal retraction (Berg and Kleinfeld 2003). The amplitude of whisking from maximum protraction to maximum retraction is ~50° (group 5 in Table 2.2). Following facial nerve transection alone, whiskers become caudally oriented and exhibit denervation-induced tremor or are motionless (Semba and Komisaruk 1984).

3.1.3 For All Treatments (Sensory, Mechanical, and Sensory + Mechanical Stimulation) Functional Outcome Correlates with Quality of Target Muscle Reinnervation

Persistent polyneuronal innervation of muscle fibers has been considered as a classic factor limiting recovery (Schroder 1968; Friede and Bischhausen 1980; Gorio et al. 1983; Grimby et al. 1989; Trojan et al. 1991; Barry and Ribchester 1995; Tam and Gordon 2003). Formation of end plates innervated by different

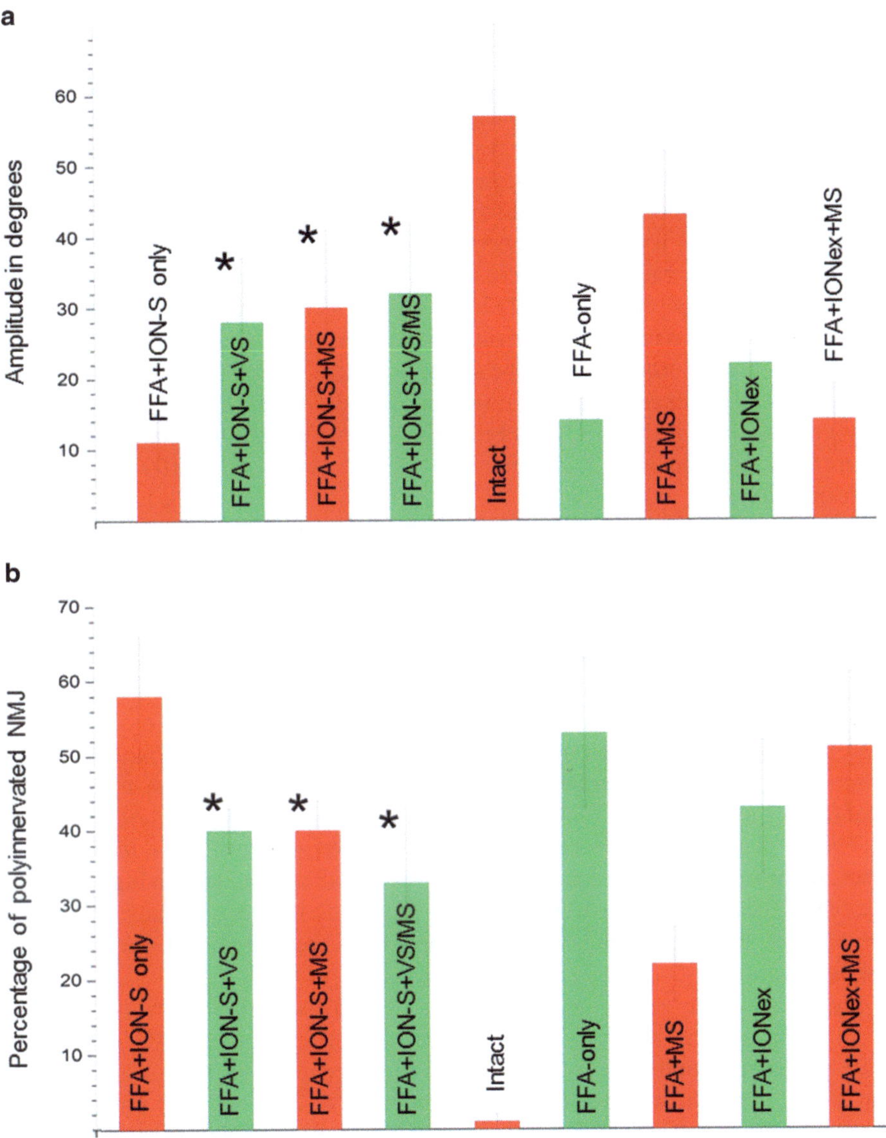

Fig. 3.1 (**a**, **b**) Mean amplitudes of vibrissal whisking (**a**) and percentage of polyinnervated neuromuscular junctions (**b**) at 4 months after surgery in the experimental groups (1–4). Data for groups 5–9 are adopted from Angelov et al. (2007) and Pavlov et al. (2008) because they were achieved in an absolutely identical manner; anyway, they were not included in the analysis. Note that the amplitude and the polyinnervation degree in the untreated postoperatively group 1 (FFA + ION-S only) remained significantly worse than those in the stimulated groups (FFA + ION-S + VS, FFA + ION-S + MS, FFA + ION-S + VS/MS). ION-ex is excision of the infraorbital nerve. Abbreviations are in Table 2.1

neurons on individual muscle fibers is a transient phenomenon during normal development. In contrast, following reinnervation polyneuronal innervation persists for longtime periods after establishment of nerve–muscle contacts (Esslen 1960; Mackinnon et al. 1991; Jergovic et al. 2001; Grant et al. 2002; Ijkema-Paassen et al. 2002), and general reasoning suggests that the performance of a muscle fiber controlled by two or more asynchronously firing motoneurons cannot be physiologically advantageous (Guntinas-Lichius et al. 2005b).

Another form of aberrant reinnervation is that end plates of different muscle fibers are serially approached and contacted by single axonal collaterals (Rich and Lichtman 1989; Son et al. 1996; Trachtenberg and Thompson 1996). Normally, the wiring pattern within a motor unit is parallel, that is, each terminal axonal branch of a motoneuron supplies one single end plate, which enables synchronous contraction of the muscle fibers. Formation of serial synapses leads to asynchronous contractions evident in EMG recordings from single motor units in patients (Montserrat and Benito 1988; Sumner 1990; Fu and Gordon 1997; Guntinas-Lichius 2004).

The degree of intramuscular sprouting and polyneuronal innervation can be manipulated, for instance, by muscle activity imposed artificially during the phase of synaptic formation and consolidation (Brown et al. 1977; Brown and Holland 1979; Al-Majed et al. 2000; Tam et al. 2001; Brushart et al. 2002; Love et al. 2003; Angelov et al. 2007), but the precise mechanisms remain to be elucidated.

Qualitative examination of the levator labii superioris muscle showed that, compared to animals which did not receive any treatment, those receiving VS, MS, or VS/MS had larger muscle fiber diameters and a higher incidence of intramuscular axonal branches. Our qualitative observations were matched by quantitative assessments of vibrissal function and the degree of polyinnervation.

Thus, it is evident that lesioned animals with no treatment (FFA + ION-S only) when compared to intact animals had poor function (Table 2.2) and a high percentage of polyinnervated end plates (58 $\pm$ 8.3 % vs. 0 % in intact animals; Table 2.3; see also the first bar in Fig. 3.1b for graphical representation of the degree of polyinnervation).

By contrast, rats receiving VS, MS, or VS/MS had better vibrissal function (Table 2.2) that was associated with significantly reduced ($p > 0.05$) motor end plate polyinnervation (Table 2.3; see also the second, third, and fourth bars in Fig. 3.1b for graphical representations) compared to no treatment. Furthermore, there was no difference in the extent of polyinnervation between the different treatments.

3.1.4 Numbers of Synaptophysin-Positive Axon Boutons in the Facial Nucleus Are Unaffected, Regardless of the Treatment

The total synaptophysin-positive particle covered area fraction of the intact lateral facial subnucleus was 17.29 % $\pm$ 2.00 % (Mean $\pm$ S.D., $n = 24$). In the experimental groups, the synaptophysin-positive fraction decreased and was not restored regardless of the treatment (Table 2.4, second column).

The intact facial motoneurons of the lateral facial subnucleus (identified by retrograde labeling with FB) had a mean of 22.41 ± 8.54 synaptic profiles on their surface (Mean $\pm$ S.D., $n = 24$). The linear synaptic density per millimeter length, a measure which corrects for perikaryal size, was 164.90 ± 33.00. Numbers of synapses per motoneuron (Table 2.4, third column) and synaptic density (Table 2.4, fourth column) decreased to a similar extent with no difference regarding the treatment in the experimental groups.

The confocal laser scanning microscope (CLSM) analysis of the sections showed similar results (Table 2.4; fifth and sixth columns).

3.1.5 No Neuronal Loss in the Trigeminal Ganglion After ION Lesion

3.1.5.1 Intact Ganglion

Following injection of 100 µl 1 % Fast Blue (FB) into the intact left whisker pad, we counted $13,980 \pm 1,709$ (mean $\pm$ S.D., $n = 24$ rats) retrogradely labeled trigeminal pseudounipolar cells, the dendrites of which projected into the intact ION (Fig. 2.6b).

No detectable loss of trigeminal neurons was found after transection and suture of the infraorbital nerve (Fig. 2.6c). The trigeminal ganglion of operated, non-stimulated rats (group 1, FFA + ION-S only) contained $13,924 \pm 4,785$ retrogradely labeled cells. Obviously the trigeminal ganglion cells not only survived the axotomy but also were able to regrow their processes to the whisker pad 4 months thereafter.

Both kinds of postoperative stimulation caused no significant alterations: In animals subjected to vibrissal stimulation (group 2, FFA + ION-S + VS), we counted $14,910 \pm 2,403$; in rats which underwent manual stimulation (group 3, FFA + ION-S + MS), $14,211 \pm 2,016$; and in animals that received combined postoperative treatment (group 4, FFA + ION-S + VS/MS), $15,138 \pm 2,442$ retrogradely labeled trigeminal ganglion cells.

3.2 Intensive Indirect Stimulation of the Trigeminal Afferents by Excision of the Contralateral ION Attenuates the Degree of Collateral Axonal Branching and Improves the Accuracy of Muscle Reinnervation

3.2.1 Reduced Degree of Collateral Axonal Branching as Determined by Application of Two Fluorescent Dyes on the Transected Superior and Inferior Buccolabial Rami of the Buccal Facial Branch

3.2.1.1 Behavioral Observations

The ability to observe postoperative vibrissae paralysis and the gradual recovery of rhythmical whisking is a major advantage of the "facial nerve model." This model allowed us to observe the rapidly improving motor performance of the mystacial vibrissae after a combined facial–trigeminal (contralateral) lesion.

Under *normal physiological conditions*, the mystacial vibrissae of the rat are erect with anterior orientation. Their simultaneous sweeps known as "whisking" or "sniffing" (Semba et al. 1980; Welker 1964) occur 5–11 times per second (Carvell and Simons 1990; Carvell et al. 1991; Komisaruk 1970). The key movements of this motor activity are the protraction and retraction of the vibrissal hairs by the piloerector muscles. After ligation of the marginal mandibular branch, all vibrissal muscles are innervated solely by the buccal branch of the facial nerve (Dörfl 1982, 1985). Accordingly, the time course of whisking recovery was prolonged in the present study and retarded in comparison with our earlier work, where the marginal mandibular branch was not eliminated (Angelov et al. 1999).

Following *buccal–buccal anastomosis (BBA)*, that is, transection and suture of the buccal facial nerve, the vibrissae dropped and acquired caudal orientation. At 10–14 days post operation (DPO), the vibrissae "rose" again to the level of the mouth and acquired a posterior orientation. Initial signs of *restoration of rhythmical whisking occurred at 21–28 DPO.*

Following *BBA plus excision of the ipsilateral infraorbital nerve*, the vibrissae "rose" to the level of the mouth and acquired a posterior orientation also at 10–14 DPO. However, in all operated rats, there was *no sign of recovery of rhythmical whisking until 28 DPO, the end of the observation period.*

Following *BBA plus excision of the contralateral infraorbital nerve*, the vibrissae "rose" to the level of the mouth and initiated movements at 3 DPO; initial signs of *restoration of rhythmical whisking occurred at 7–10 DPO.*

The observation that the alterations induced by excision of the contralateral infraorbital nerve provided not only the most rapid but also the qualitatively best recovery of function (rhythmical whisking of the vibrissae) was so unexpected and exciting that we checked it in an electrophysiological and morphological way.

3.2.1.2   Electrophysiological Measurements

Stimulation of the normal buccal nerve generated a whisker pad muscle potential of 28.4 ± 1.0 mV (mean $\pm$ S.D.; $n = 8$ rats). The amplitude was constant over the time course: 27.5 ± 1.0 mV 7 days later, 27.5 ± 0.6 mV 28 days later, and 28.6 ± 0.8 mV 42 days after the first measurement.

Seven days after surgery, the amplitude of the evoked electromyography waveform dropped to 12.5 ± 0.3 mV (45 % of the normal value) after unilateral BBA, to 11.4 ± 0.5 mV (42 %) after BBA plus excision of the ipsilateral infraorbital nerve, and to 19.8 ± 0.3 mV (72 %) after BBA plus excision of the contralateral infraorbital nerve.

Twenty-eight days after surgery, the amplitude rose to 20.4 ± 0.2 mV (74 %) after unilateral BBA, to 19.49 ± 0.6 mV (70 %) after BBA plus excision of the ipsilateral infraorbital nerve, and to 25.85 ± 0.2 mV (94 %) after BBA plus excision of the contralateral infraorbital nerve.

Finally, at 42 DPO, 27.4 ± 0.3 mV (96 %) could be measured after unilateral BBA, 28.4 ± 0.2 mV (99 %) after BBA plus excision of the ipsilateral infraorbital nerve, and 28.23 ± 0.2 mV (99 %) after BBA plus excision of the contralateral infraorbital nerve.

In comparison to the control group, a statistically significant decrease in amplitude was found at 7 and 28 days after surgery for all experimental groups. In comparison to the control group, a statistically significant decrease in amplitude was found at 42 days after surgery only for unilateral BBA.

In conclusion, the amplitude recovered to nearly normal values after any type of operation but significantly faster after BBA plus excision of the contralateral infraorbital nerve.

3.2.1.3   Estimation of Axonal Regrowth and Branching

Normal Values in Intact Rats

Application of crystals of FluoroGold to the superior and crystals of DiI to the inferior buccolabial nerve yielded $1,724 \pm 375$ FG- and 134 ± 125 DiI-labeled motoneurons (mean $\pm$ S.D., $n = 4$). All retrogradely labeled cells (total of $1,858 \pm 424$) were localized exclusively in the lateral facial subnucleus. However, whereas the FG-labeled cells were found in its ventrolateral portion, the DiI-labeled cells were observed in the dorsomedial portion of the lateral facial subnucleus. No double-labeled motoneurons were observed (data not shown; see, however, Fig. 2.8a, c, e, demonstrating the same finding in the intact contralateral facial nucleus of rats which underwent single or combined surgery on the buccal nerve).

Alternatively, application of DiI to the superior and FG to the inferior buccolabial nerve yielded $1,937 \pm 156$ DiI- and 94 ± 30 FG-labeled motoneurons (mean $\pm$ S.D., $n = 4$). All DiI-labeled cells were located in the ventrolateral

Table 3.1 Projection pattern of facial motoneurons after facial nerve lesion in rats

Surgery	Superior buccolabial nerve	Inferior buccolabial nerve	Superior + inferior buccolabial nerves	Buccal facial nerve
Intact rats (Group 1)	1,746 ± 375 91 %	174 ± 91 9 %	0	1,920 ± 288 100 %
28 days after buccal-bucal anastomosis (BBA) only (Group 2)	838 ± 499* 56 %	312 ± 142[§] 21 %	342 ± 352* 23 %	1,491 ± 604 100 %
112 days after BBA only (Group 2a)	1,004 ± 393 63 %	416 ± 288 26 %	172 ± 84 11 %	1,591 ± 484 100 %
28 days after BBA + excision ipsilateral ION, (Group 3)	860 ± 439* 48 %	678 ± 426*[§] 39 %	237 ± 249 13 %	1,776 ± 476 100 %
112 days after BBA + excision ipsilateral ION (Group 3a)	1,064 ± 422 60 %	511 ± 204 29 %	198 ± 70 11 %	1,772 ± 375 100 %
28 days after BBA + excision contralateral ION (Group 4)	1,271 ± 352 69 %	418 ± 247 23 %	164 ± 115 8 %	1,855 ± 581 100 %
112 days after BBA + excision contralateral ION (Group 4a)	1,266 ± 263 76 %	312 ± 238 19 %	89 ± 54 5 %	1,667 ± 243 100 %

Number of retrogradely labeled motoneurons, the axons of which project within the superior, inferior, or both buccolabial nerves in intact rats, in rats 28 and 112 days after unilateral buccal-buccal anastomosis (BBA), after BBA plus excision of the ipsilateral infraorbital nerve (ION), and after BBA plus excision of the contralateral ION. Means ± S.D; $n = 8$ rats in group 1, $n = 12$ rats groups 2–4, and $n = 6$ rats in groups 2a–4a
*$P \leq 0.05$ for the comparison with group 1 (intact rats)
[§]$P \leq 0.05$ for the comparison between group 2 (BBA only) and group 3 (BBA plus excision of ipsilateral infraorbital nerve)

portion and all FG-labeled cells were observed in the dorsomedial portion of the lateral facial subnucleus, which contained 2,031 ± 178 retrogradely labeled motoneurons. No double-labeled motoneurons were observed.

The statistical evaluation showed that these numbers are practically identical (*t*-test for unpaired data), and no difference in the labeling efficiency of FG and DiI was found in this experimental setup. Thus, after pooling these data, the lateral facial subnucleus was shown to contain 1,920 ± 288 motoneurons. About 91 % of these motoneurons (1,747 ± 375) project into the superior and 9 % (174 ± 92) into the inferior buccolabial nerve (Table 3.1). There were no motoneurons projecting through both buccolabial nerves.

Buccal–Buccal Anastomosis

Neuron labeling at 28 days after buccal–buccal anastomosis (BBA) showed that all retrogradely labeled neurons were localized in the lateral facial subnucleus. The quantitative estimates revealed no neuronal loss (Table 3.1). A myotopic organization of this subnucleus into a ventrolateral (for the superior buccolabial nerve) and a dorsomedial (for the inferior buccolabial nerve) portion was, however, no more evident (Fig. 2.8b).

Accordingly, the number of motoneurons whose axons or axonal branches projected into the superior buccolabial nerve was lower than that in intact rats: Only about 56 % of all neurons in the lateral facial subnucleus projected into the superior buccolabial nerve. On the contrary, due to the misguided growth of axons into wrong fascicles, the number of motoneurons whose axons projected into the inferior buccolabial nerve was increased in comparison with that in intact rats: The motoneurons whose axons have regrown into the inferior buccolabial nerve comprised about 21 % of all neurons in the lateral facial subnucleus (Table 3.1). Compared to the values in intact rats, there is a statistically significant decrease in the number of motoneurons projecting through the superior buccolabial nerve.

Another major difference from unoperated animals was the presence of motoneurons containing both tracers. The only explanation to this is that these double-labeled cells (23 % of all motoneurons in the lateral facial subnucleus) regrew several (but not one single) sprouts from one transected axon, which postoperatively projected within the superior and inferior buccolabial nerves (Shawe 1954; Esslen 1960; Brushart and Mesulam 1980; Ito and Kudo 1994). This suggestion was confirmed by the neuron counts at 112 days after BBA: The portion of double-labeled motoneurons in group 2a was reduced to 11 % (Table 2.1) and confirmed earlier results that most of the supernumerary sprouts have been pruned off (Shawe 1954; Brushart 1993; Mackinnon et al. 1991).

BBA Plus Excision of the Ipsilateral Infraorbital Nerve

Four weeks after BBA plus excision of the ipsilateral infraorbital nerve, all retro-gradely labeled neurons ($1,776 \pm 476$) were localized in the lateral facial subnucleus but with no myotopic organization (Fig. 2.8d).

The motoneurons whose axons projected into the superior buccolabial nerve comprised about 48 % of all neurons in the lateral facial subnucleus. Additionally, the portion of neurons whose axons projected postoperatively into the inferior buccolabial nerve rose to about 39 % (Table 2.1). Double-labeled neurons, projecting into both fascicles of the buccal nerve, comprised about 13 % of all motoneurons in the lateral facial subnucleus.

Compared to the values in intact rats, there is a statistically significant decrease in the number of motoneurons projecting within the superior nerve and a significant increase in the number of motoneurons projecting to the inferior nerve.

Compared to BBA alone, there is no statistically significant change in the number of motoneurons projecting within the superior nerve. There is, however, a significant increase in the number of motoneurons projecting into the inferior nerve and a statistically significant decrease in the number of motoneurons projecting to both fascicles.

The neuron counts at 112 days after surgery (group 3a) showed no significant differences from the values obtained at 4 weeks after surgery (Table 3.1).

BBA Plus Excision of the Contralateral Infraorbital Nerve

Neither obvious neuronal loss nor involvement of other subnuclei of the facial nucleus occurred after BBA plus blockade of the contralateral trigeminal input: All retrogradely labeled neurons (1,855 ± 581) were localized in the lateral facial subnucleus, but the myotopic organization was lost (Fig. 2.8f).

The number of motoneurons whose axons or axonal branches projected into the superior buccolabial nerve comprised 69 % of all neurons in the lateral facial subnucleus. The mean number of neurons whose axons projected into the inferior buccolabial nerve decreased to 23 % (Table 3.1) and the mean number of double-labeled neurons to about 8 % of all motoneurons in the lateral facial subnucleus (Table 3.1).

Compared to the values in intact rats, there are no statistically significant changes in the number of motoneurons projecting into the superior nerve, the inferior nerve, and into both nerves.

Compared to BBA alone, there is a statistically significant increase in the number of motoneurons projecting within the superior nerve, no significant change in the number of cells projecting into the inferior nerve, and no statistically significant decrease in the number of motoneurons projecting to both nerves.

Compared to BBA plus excision of the ipsilateral infraorbital nerve, there is a statistically significant increase in the number of motoneurons projecting within the superior nerve, a statistically significant decrease in the number of motoneurons projecting into the inferior nerve, and a statistically significant decrease in the number of motoneurons projecting to both nerves.

The neuron counts at 112 days after surgery showed significant reduction in the number of double-labeled cells only in group 4a (Table 3.1).

Taken together, the results from retrograde neuron labeling showed that, when compared to transection and suture alone (BBA), the lesion of the infraorbital nerve reduced the branching (or enhanced the elimination of branches) from transected facial axons. Surprisingly, the best prerequisites for recovery of function were detected after lesioning the trigeminal ganglion neurons contralateral to the lesioned facial nucleus. Furthermore, it turned out that injuring the ipsilateral trigeminal ganglion neurons yielded results, which were even worse than those after BBA alone. Since we failed to provide a sound explanation, we decided to check anatomically whether the excision of the contralateral infraorbital nerve accelerates the elongation of transected facial axons.

3.2.1.4 Estimation of Axonal Elongation Rate

As already pointed in "Chap. 2," 24 rats were used for this experimental set. In the course of observation and counting, it became clear that five rats had a small ramus from the marginal mandibular nerve projecting to the vibrissal area. This was detected by the presence of a rather compact group of FB-labeled motoneurons in the intermediate facial subnucleus. These five animals were excluded from statistical analysis.

Injection of 100 µl 1 % FB into the whisker pad of normal intact rats labeled 1,602 ± 96 motoneurons (mean ± S.D., $n = 6$ rats) localized exclusively in the lateral facial subnucleus.

Three days after tracer injection and 4 days after *BBA only*, we counted 1,503 ± 102 retrogradely labeled motoneurons in the intact lateral facial nucleus on the control side. On the side where BBA was performed, there were 308 ± 57 (range from 222 to 363) retrogradely labeled motoneurons (mean ± S.D.; $n = 5$ rats). All of them were localized in the lateral facial subnucleus. This value is significantly lower than the number established in intact control animals.

Three days after tracer injection and 4 days after *BBA plus excision of the ipsilateral infraorbital nerve*, there were 1,579 ± 114 retrogradely labeled motoneurons in the intact lateral facial nucleus on the control side. On the side of combined surgery, we counted 218 ± 130 (range from 89 to 432) motoneurons projecting into the muscles of the whisker pad ($n = 4$ rats). Compared to the mean value in unoperated control animals and to the mean value in animals which underwent BBA only, this number is significantly lower.

Three days after tracer injection and 4 days after *BBA plus excision of the contralateral infraorbital* nerve, we counted 1,605 ± 141 motoneurons in the intact lateral facial nucleus on the control side. On the side of combined surgery, there were 434 ± 74 (range from 357 to 501) motoneurons whose axons had reached the whisker pad muscles ($n = 4$ rats). Compared to the mean value in unoperated control animals, this number is significantly lower. Compared to the mean value in animals which underwent BBA only, this value is not significantly higher. Compared to the mean value in animals which underwent BBA plus excision of the ipsilateral infraorbital nerve, this value is significantly higher.

Accordingly, this part of our results did not provide evidence for an increased rate of facial axon elongation after combined facial–trigeminal injury: The combination of facial axotomy plus excision of the contralateral infraorbital nerve turned out to be superior to BBA plus excision of the ipsilateral infraorbital nerve, but not to BBA only. The combination of intramuscular tracer injection and subsequent counts of retrogradely labeled motoneurons has two main advantages (1) It determines the source and amount of normal nerve supply, and (2) it allows to establish the time course of long-term muscle reinnervation. However, counts of retrogradely labeled motoneurons might be of relatively low precision when employed after nerve lesions performed close to the target muscle. Whereas it is established that even the earliest axonal branches after axotomy are capable of tracer incorporation (cf. Sparrow and Kiernan 1979; Olsson 1980), little is known what portion of these branches succeeds to establish neuromuscular junctions later. As in our model the distance between nerve transection site and target musculature is about 12 mm, the method chosen appears to be fairly crude. Pilot experiments to visualize the advancement of outgrowing axons from the site of transection are going on.

3.2.2 *Improved Accuracy of Reinnervation as Established by Means of Intramuscular Injections of Fluorescent Dyes and Electrophysiological Measurements*

3.2.2.1 Estimation of Axonal Regrowth and Branching

Single Labeling in Intact Rats

The facial motoneurons of intact rats were labeled by either FG or FB to compare the *labeling efficiency* of these two retrograde fluorescent tracers. Injection of 100 µl 1 % FG into the whisker pad (right or left) labeled the perikarya of $1{,}281 \pm 87$ facial motoneurons (mean $\pm$ S.D., $n = 8$ rats). Injection of 100 µl 1 % FB into the whisker pad (right or left) labeled the perikarya of $1{,}302 \pm 96$ motoneurons (mean $\pm$ S.D., $n = 8$ rats). Thus, there appeared to be no difference in the labeling efficiency of FG and FB in our experimental system. With both tracers, all labeled motoneurons were localized exclusively in the lateral facial subnucleus, which is in agreement with the previously described myotopic organization of the facial nucleus in normal rats (Aldskogius and Thomander 1986; Angelov et al. 1996).

Sequential Labeling

As a second methodological control, we tested whether the sequential injection of FG and FB in the selected muscle target could provide a reliable distinction between the FG (preoperatively labeled), FB (postoperatively labeled), and FG + FB (double-labeled) neurons. Employing the custom-made selective filter sets for FG and FB, we could differentiate and save separate digital images of FG (orange-red) and FB (blue) motoneuronal profiles. Our results showed that in intact rats the portion of double-labeled (FG + FB) motoneurons is about 90 % (Table 3.2, first row), which is reasonably close to the theoretical expectation of 100 % double labeling.

Pre- and Postoperative Labeling

A wealth of experimental studies has shown that intramuscular injection of a tracer, for example, horseradish peroxidase (HRP), and counts of retrogradely labeled motoneurons provide a powerful and reliable method to evaluate muscle reinnervation (Angelov et al. 1993, 1995, 1997; Dohm et al. 2000; Guntinas-Lichius et al. 2000; Son et al. 1996). Unfortunately, quantitative estimation after single tracing with HRP provides information solely about the reinnervation status of a muscle target since the one-tracer approach cannot elucidate the relationship between pre- and postoperative innervation of target muscles. On the other hand, the sequential

Table 3.2 Mean numbers and standard deviations (S.D.) of retrogradely labeled motoneurons following preoperative intramuscular application of 100 µl 1 % FG and postoperative injection of 100 µl 1 % FB in the whisker pad of (1) intact rats and in rats 28 days after (2) buccal–buccal anastomosis (BBA), (3) BBA plus excision of the ipsilateral infraorbital nerve (ION), and (4) after BBA plus excision of the contralateral ION

	Intact side			Operated side		
Surgery	FG-labeled original pool	FB-labeled original pool	FG + FB-labeled original pool	FG-labeled original pool	FB-labeled postoperative pool	FG + FB-labeled accurately regrown
Intact rats ($n = 6$)	$1,425 \pm 52$	$1,413 \pm 50$	$1,290 \pm 87$ 91 %	$1,264 \pm 114$	$1,305 \pm 137$	$1,169 \pm 224$ 90 %
BBA ($n = 9$)	$1,388 \pm 112$	$1,427 \pm 176$	$1,272 \pm 303$ 89 %	$1,119 \pm 109$	$1,456 \pm 132$	398 ± 80 27 %
BBA + excision of ipsilat ION ($n = 9$)	$1,177 \pm 94$	$1,182 \pm 112$	$1,058 \pm 179$ 90 %	$1,147 \pm 95$	$1,362 \pm 162$	436 ± 68 32 %
BBA + excision of contralat ION ($n = 9$)	$1,256 \pm 67$	$1,301 \pm 82$	$1,156 \pm 100$ 89 %	$1,245 \pm 76$	$1,406 \pm 81$	580 ± 63 41 %

application of two tracers has been repeatedly used (Brushart et al. 1998; Harsh et al. 1991; Madison et al. 1996; Molander and Aldskogius 1992). The sequential application of double retrograde labeling should allow an optimal evaluation of pre- and postoperative distribution of motoneurons in the same animal, avoiding counting errors due to interindividual variability. Our own experience shows that the best combination of fluorescent retrograde tracers to study the accuracy of post-transectional muscle reinnervation is a preoperative labeling of the original moto-neuronal pool by an injection of 1 % FG into the muscle target, followed by a postoperative labeling of all motoneurons innervating the same target after surgery by an injection of 1 % FB (Popratiloff et al. 2001). All numerical values of neurons labeled with FG, FB, and FG + FB are presented in Table 3.2.

Contralateral Intact Facial Nucleus

All retrogradely labeled neurons (FG, FB, and FG + FB) were localized exclusively in the lateral facial subnucleus (Fig. 2.9a–c). Thus, on the side of the brain stem contralateral to BBA, there were no detectable differences among all three types of operations either in location or in numbers of FG-, FB-, and FG + FB-labeled neurons (one-way ANOVA; no significance).

Buccal–Buccal Anastomosis (Group BBA)

Neuronal labeling 28 days after BBA showed that all FG-labeled neurons were located in the lateral facial subnucleus (Fig. 2.9d). However, the distribution pattern of the FB-labeled neurons innervating the whisker pad musculature after BBA was changed. We observed "ectopic" neurons projecting to the whisker pad musculature, which were located in the intermediate facial subnucleus (Fig. 2.9e, f).

The quantitative estimates revealed no postoperative loss of neurons (Table 3.2, operated side, column "FB-labeled"). However, only 398 ± 80 of these FB-labeled motoneurons were double labeled (colored in pink to bright purple in Fig. 2.9f), that is, only 27 % of them belonged to the original motoneuron pool; the rest were ectopic nerve cells.

The innervation of the whisker pad by "ectopic" motoneurons (Fig. 2.9e, f, h, i, k, l) localized in the dorsal, intermediate, medial, and ventromedial facial subnuclei is a well-known phenomenon after transection of the facial nerve (Angelov et al. 1993, 1996; Streppel et al. 1998). In intact rats, the motoneurons in these subnuclei send their axons exclusively along the zygomatic, marginal mandibular, posterior auricular, and cervical branches of the facial nerve, respectively (Semba and Egger 1986). Whether the post-transectional misguidance of twin axons stemming from the intact facial branches occurred in the periphery as consequence of Schwann cell bridges (Love and Thompson 1999) or within the facial nerve trunk remains to be elucidated.

BBA Plus Excision of the Ipsilateral Infraorbital Nerve (Group BBA + Ipsi-ION-ex)

All FG-labeled neurons were located in the lateral facial subnucleus (Fig. 2.9g). FB-labeled perikarya were found in the lateral, medial, and intermediate facial subnuclei (Fig. 2.9h, i). Four weeks after BBA and excision of the ipsilateral infraorbital nerve, we counted 436 ± 68 double-labeled (pink–purple) motoneurons (Table 3.2), that is, only about 32 % of the motoneurons which innervated the whisker pad post surgery belonged to the original pool. Compared to BBA alone, there is no statistically significant increase in the number of double-labeled motoneurons.

BBA Plus Excision of the Contralateral Infraorbital Nerve (Group BBA + Contra-ION-ex)

Apart from preventing a transsagittal sprouting from the contralateral ION (Baumel 1974; Banfai 1976), the excision of the contralateral ION served to prove whether the removal of the contralateral vibrissae input to the axotomized facial motoneurons would attenuate the misguidance of regrowing facial sprouts.

All FG-labeled neurons were located in the lateral facial subnucleus (Fig. 2.9j). These cells appeared larger than the shrunken perikarya observed in the previous groups (compare Fig. 2.9j with g and d). FB-labeled motoneurons were also found in the intermediate facial subnucleus (Fig. 2.9k, l). Four weeks after BBA plus excision of the contralateral infraorbital nerve, we counted 580 ± 63 double-labeled motoneurons (Table 3.2), that is, about 41 % of the motoneurons which innervated the whisker pad muscles after this type of combined surgery sent an axon or an axonal branch to their original target. Compared to BBA alone and to BBA plus excision of the ipsilateral infraorbital nerve, there is a statistically significant change in the number of double-labeled motoneurons ($p = 0.01$).

Altogether, the results from retrograde neuron labeling show that, when compared to transection and suture alone (BBA), or to BBA plus excision of the ipsilateral infraorbital nerve, the lesion of the contralateral infraorbital nerve resulted in less aberrant reinnervation, which correlates well with the improved motor function found after this operation.

3.2.2.2   Electrophysiological Measurements of the Compound Muscle Action Potential

Unoperated Control Animals

In normal control animals, the suprathreshold stimulation of the buccal branch resulted in an obvious but variable sweep of the mystacial hairs. Following a supramaximal stimulation, the movements of the vibrissae were synchronized.

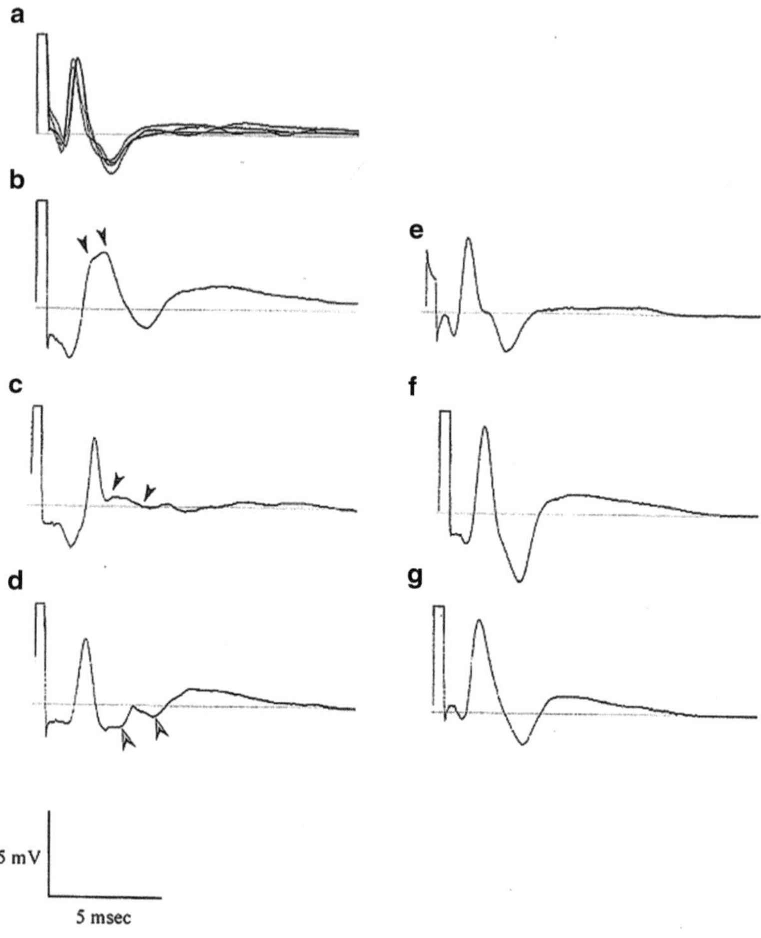

Fig. 3.2 Traces showing compound muscle action potential (CMAP) recorded from the whisker pad after supramaximal stimulation to the buccal facial branch. (**a**) Normal interindividual variability of CMAP demonstrated by four superimposed traces from four unoperated control animals. Recordings were made with a negative silver wire electrode inserted into the whisker pad. Typically, the CMAP consists of a faster negative wave, due to summation of depolarizations from numerous muscle fibers. This wave is followed by a positive wave of much smaller amplitude, resulting from the summation of the hyperpolarizations, and by a not so prominent and slow negative deflection due to the summation of afterhyperpolarizations of the muscle fibers. (**b**) CMAP of identical amplitude, but prolonged duration at 28 days after BBA. The prolonged duration is due to the disintegration of CMAP's peaks (*arrowheads*). (**c**) A typical trace of CMAP taken from rats which underwent BBA plus excision of the ipsilateral ION (BBA + ipsi ION). The fast negative wave is followed by smaller and slower depolarizations (*arrowheads*) superimposed to the hyperpolarization wave. The overall duration of the entire complex is shorter than in the group BBA. (**d**) Trace obtained from rats with BBA plus excision of the contralateral ION (BBA + contra ION). The fast depolarization is followed by a positive hyperpolarization with an occasional negative activity superimposed to the positive wave. This negative activity, however, does not exceed the zero line (*open arrowheads*). (**e–g**) Traces of CMAP recorded from the left whisker pad of groups with BBA, with BBA + ipsi ION, and with BBA + contra ION,

The threshold for achieving maximal response was low (1.5–2.5 mA). This maximal response, termed the compound muscle action potential (CMAP), had a short latency of 1.55 ± 0.09 ms; ($n = 5$). However, both latency and threshold were dependent on slight differences in electrode position, so they were disregarded for statistical analysis.

The CMAP consisted of a fast 3-phasic positive–negative–positive wave, followed by a slow and not so prominent negative wave (Fig. 3.2a). With the recording procedure used here, the faster negative wave reflected the overlapping depolarizations of all innervated muscle fibers. The short duration (1.10 ± 0.14 ms; $n = 5$ rats) of the negative wave was due to the similar conductance velocities and excitability of the involved fibers. The peak amplitude was estimated to be 4.53 ± 0.55 mV ($n = 5$). The Kolmogorov–Smirnov one-sample test revealed that the distributions for both duration and amplitude met the criteria for normal distribution (asymmetric significance: duration—0.89; amplitude—0.76), which indicated a random sampling procedure.

Buccal-to-Buccal Anastomosis (Group BBA)

In all animals of this group, the latency, the threshold stimuli for achieving a CMAP, and the stimuli for generating maximal CMAP were increased. The mean duration of CMAP was obviously increased (Fig. 3.2b) and was estimated to be 1.82 ± 0.53 ms ($n = 5$ rats). As shown in Fig. 3.2b, this increased duration was due to the prolonged time of the rising phase of the CMAP, which consisted of a fast and a slower component. This was in contrast to the CMAP in the controls, where only one a single fast rising phase was present. This phenomenon most likely reflects the variety of the conductance velocity in the regenerating buccal axons as well as the asynchronous depolarization of the whisker pad muscles from synaptic input. The t-test for unpaired samples revealed that the mean duration was significantly different from the value obtained in the controls (Levene test: Sig < 0.05; 2 tailed significance, equal variances not assumed $= 0.035 < 0.05$). The mean amplitude was lower than the mean value in control animals (Fig. 3.2b), but this difference was not significant (3.86 ± 0.63; $n = 5$).

BBA Plus Excision of the Ipsilateral Infraorbital Nerve (Group BBA + Ipsi-ION-ex)

Like the animals in the BBA group, the latency, threshold stimuli, and stimuli for generating maximal CMAP were increased (Fig. 3.2c). The CMAP consisted typically of a fast negative wave, the mean duration of which was shorter than in

Fig. 3.2 (continued) respectively. There are no obvious changes in the CMAP's shape, duration, and peak, when compared with those obtained from the control animals (see trace (**a**))

the BBA group. However, this mean duration (1.36 ± 0.12; $n = 5$) was significantly longer than the respective value in control animals (Levene test significance > 0.05; equal variances assumed; t-test significance < 0.05). The amplitude of the CMAP (4.27 ± 0.98; $n = 5$) did not differ significantly from that obtained in control animals (Levene test significance > 0.05; two independent sample t-test significance < 0.05; 95 % confidence interval of the difference contained "0").

BBA Plus Excision of the Contralateral Infraorbital Nerve (Group BBA + Contra-ION-ex)

Like the animals in the first experimental group, the latency, the threshold stimuli, and the stimuli for generating maximal CMAP were increased (Fig. 3.2d). The CMAP consisted of a fast wave, the mean duration of which was shorter than that measured in the BBA and BBA + ipsi ION groups. This estimated mean duration (1.08 ± 0.14 ms; $n = 5$) was not significantly different from the control animals (Levene test significance > 0.05; two independent sample equal variance t-test significance > 0.05). The mean amplitude (3.81 ± 0.50; $n = 5$) also did not differ significantly from the value measured in the control animals (Levene significance > 0.05; two independent sample t-test equal variance presumed significance > 0.05).

In conclusion, the reduced mean duration of CMAP in the BBA + contra-ION-ex group reflects synchronized contractions of the whisker pad muscles, which closely resemble those in control animals.

3.3 Direct Stimulation of the Trigeminal and Facial Nerves by Massage of the Vibrissal Muscles Improves the Quality of Target Reinnervation and Promotes Full Recovery of Whisking Function

3.3.1 Analysis of Vibrissae Motor Performance During Exploration

We first examined the effects of manual stimulation on the recovery of function. Video analysis of vibrissal motion allows rapid assessment of motor function (Guntinas-Lichius et al. 2002; Tomov et al. 2002). Normal animals explore the environment by coordinated sweeps of individual vibrissae ("whisking") with a frequency of about 6 Hz (Semba et al. 1980; Bermejo et al. 1996; Komisaruk 1970; Carvell and Simons 1990). Vibrissal movements are characterized by an active protraction rostrally via muscle contraction and, as recently described, by an active retraction caudally (Berg and Kleinfeld 2003). The amplitude of the movement from maximum protraction to maximum retraction is about 50° (Fig. 3.3a, d; Table 2.6).

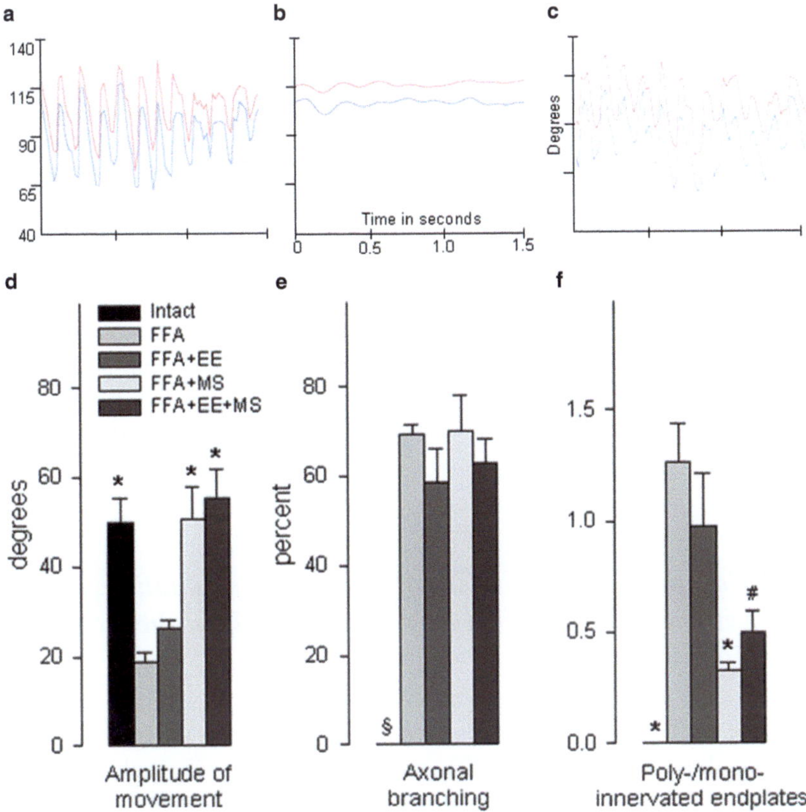

Fig. 3.3 (**a–c**) Recordings of vibrissae movements in an intact rat (**a**) non-stimulated (**b**) and manually stimulated (**c**) animal 2 months after nerve repair. The graphs show movements of two large C-row vibrissae (*red* and *blue*). The parallel course of the curves indicates the synchronous movements of the two vibrissae. (**d–f**) Major parameters evaluated in intact rats (*intact*), rats subjected to FFA without mechanical stimulation (*FFA*), FFA plus enriched environment (*FFA + EE*), FFA plus mechanical stimulation (*FFA + MS*), or FFA plus combined treatment (*FFA + EE + MS*) for 2 months after facial–facial anastomosis (FFA). Mechanical stimulation (*FFA + MS*), but not enriched environmental housing (*FFA + EE*), leads to full recovery of the amplitude of vibrissal motion (**d**). The extent of axonal branching in the facial nerve trunk is not influenced by any of the stimulation protocols (**e**). The ratio of polyinnervated to monoinnervated motor end plates is strongly reduced as a result of manual, but not environmental, stimulation (**f**). The index of axonal branching represents the ratio of motoneurons projecting branched axons into the zygomatic and the buccal or mandibular branch to motoneurons with unbranched axons innervating the zygomatic branch only (see Table 2.7). Note that the values of the two structural parameters shown are 0 in intact animals. Values are the mean ± SEM. Groups indicated by symbols are significantly different ($p < 0.05$, ANOVA and Tukey *post hoc* test) compared to *—groups FFA and FFA + EE; #—group FFA; §—all other groups

Recovery of vibrissal motion after facial nerve injury (1) alone, that is, without mechanical stimulation, (2) after mechanical stimulation for 1 and for 2 min only, (3) after stimulation of the contralateral vibrissal muscles, and (4) after postoperative handling was poor most likely due to inadequate muscle function during protraction and retraction (Berg and Kleinfeld 2003): The amplitude of movements and the angular velocity were reduced to less than 40 % and 23 % of the values in intact animals, respectively (Fig. 3.3b, d; Table 2.6). Frequency of whisking after nerve repair was similar to that in intact rats (see second column in Table 2.6) which may be due to the robust autonomy and capacity of the whisking pattern generator, represented mostly by neurons projecting to the facial nucleus from the brain stem (Popratiloff et al. 2001; Hattox et al. 2002; Veinante and Deschenes 2003).

Manual stimulation of the ipsilateral whiskers for 5 and for 10 min daily had a dramatic effect, resulting in a return of normal whisking as indicated by the amplitude of movement (Fig. 3.3c, d), as well as by the speed during protraction (Table 2.6). Stimulation by an enriched environment did not result in a recovery of function although combined stimulation (manual and environmental) did (Table 2.6).

3.3.2   Manual Stimulation Counteracts Posttraumatic Loss of Synaptophysin-Positive Axon Terminals in the Facial Nucleus

Analyses of the frequency distributions of pixel intensities revealed no differences among the three groups (intact, FFA only, and FFA + MS) indicating similar overall intensities of the immunofluorescence staining. The background intensities were identical from image to image around a pixel gray value of 30. Therefore, we used the total number of pixels per frame within the range (gray values 30–129) as an estimate of axon terminal density.

In animals with FFA and no MS, the mean total number of pixels was significantly lower than in intact animals ($29.2 \times 10^6 \pm 1.8 \times 10^6$ vs. $34.3 \times 10^6 \pm 2.3 \times 10^6$; $p = 0.036$, ANOVA with Bonferroni post hoc test). In animals receiving MS after FFA, the mean pixel number was similar to intact animals ($33 \times 10^6 \pm 2.6 \times 10^6$) but higher than in the control FFA group ($p = 0.007$). These results indicate that the synaptic input after FFA alone is reduced compared to normal and this loss is counteracted by MS.

3.3.3   Degree of Collateral Axonal Branching Remains Elevated Regardless of Stimulation

We used triple retrograde labeling to assess the projection patterns of motor axons from the facial nucleus through its different motor rami as well as the degree of axonal branching (Dohm et al. 2000). In normal animals, motoneurons with axons entering the zygomatic, buccal, or marginal mandibular ramus (Fig. 2.11a) were localized in the dorsal, lateral, and intermediate facial subnuclei, respectively (Hinrichsen and Watson 1984; Klein and Rhoades 1985; Aldskogius and Thomander 1986; Semba and Egger 1986). No double- or triple-labeled motoneurons were observed because intact motoneurons send only one unbranched axon to one of the facialis rami (Fig. 2.11b, Table 2.7). Thus, the index of axonal branching in the facial nerve trunk calculated from the zygomatic motoneurons was 0 % (Fig. 3.3e).

After facial nerve injury and no stimulation, myotopic organization into subnuclei was no longer observed, that is, all retrogradely labeled motoneurons were scattered throughout the facial nucleus (Fig. 2.11d). This was due to the numerous collateral branches (up to 25; Shawe 1954) emerging from a single transected axon. These persistent branches joined different facial nerve rami and transported retrogradely different fluorescent dyes to motoneurons located randomly in the facial nucleus.

In addition, double and triple labeling of motoneurons was commonly observed (Fig. 2.11d). The only explanation for this multiple labeling is that the axonal branches which projected through different facial nerve rami (e.g., zygomaticus, buccalis, or marginalis mandibulae that were instilled with different tracers) took up and transported retrogradely two or three fluorescent dyes simultaneously. This in turn led to double and triple labeling of their parent perikarya. Finally, as a result of this robust collateral branching at the site of lesion, each of the individual facial nerve rami contained axons or axonal branches of more motoneurons than in intact animals (Table 2.7), which in the periphery caused the hyperinnervation of targets (Rich and Lichtman 1989; Angelov et al. 1996; Streppel et al. 1998).

Transection and suture of the facial nerve dramatically affected the fiber composition of the different nerve rami which we traced. The total number of labeled motoneurons in FFA-treated animals increased from 2,184 $\pm$ 242 (in intact rats) up to 3,609 $\pm$ 858 (after FFA), 5,127 $\pm$ 993 (after FFA + EE + MS), 5,563 $\pm$ 841 (after FFA + EE), and 6,328 $\pm$ 745 (after FFA + MS). This very high motoneuron number was due to a significant increase in the number of motoneurons single-labeled by FluoroGold and Fast Blue, respectively (Table 2.7). The axonal sprouts which took up retrograde tracer most likely originated from motoneurons projecting under control conditions only to the posterior auricular or cervical ramus. Since these rami were not traced in the intact animals, the corresponding motoneurons were not visible and therefore not counted. After transection of the facial nerve trunk, the emerged collateral branches of axons, which originally projected into ramus auricularis posterior and ramus colli, presumably grew into one of the traced facial nerve rami (buccal or marginal mandibular).

In the present study, the estimated index of axonal branching was 70 % (Fig. 3.3e). None of the stimulation paradigms, that is, manual, environmental, or combined, had any significant influence on the projection patterns (Fig. 3.3e, Table 2.7). Thus, there was a complete lack of myotopic organization, increased total numbers of projecting motoneurons, and a consistently elevated degree of axonal branching regardless of whether the animals were subjected to any of the stimulation paradigms or not.

3.3.4 *Mechanical Stimulation Reduces the Degree of Motor End Plate Polyinnervation*

We then investigated the reinnervation pattern at the level of individual skeletal muscle fibers of the levator labii superioris muscle (Fig. 2.2a; Guntinas-Lichius et al. 2005b). Although post-lesional polyinnervation of the end plates has been claimed to be transient (Hennig and Dietrichs 1994), accumulating evidence suggests that it persists after establishment of nerve–muscle contacts (Esslen 1960; Mackinnon et al. 1991; Reynolds and Woolf 1992; Madison et al. 1999; Jergovic et al. 2001; Ijkema-Paassen et al. 2002; Grant et al. 2002; Choi and Raisman 2005), and our previous work indicates that post-lesional polyinnervation has a deleterious effect on recovery of facial motor function (Guntinas-Lichius et al. 2005b).

In intact animals, all motor end plates were monoinnervated (Fig. 2.2c). After facial nerve injury and no stimulation, 53 % were polyinnervated, that is, innervated by two or more axons (Fig. 2.2b, Table 2.8). However, manual, but not environmental, stimulation significantly reduced the degree of polyinnervated end plates (22 % and 41 %, respectively), whereas the combination of both had an intermediate effect (31 %; Table 2.8). Thus, manual stimulation reduced the ratio between poly- and monoinnervated end plates (Fig. 3.3f) by a factor of four compared to untreated rats and to a level which did not differ statistically from intact animals.

3.3.5 *Manually Stimulated Recovery of Motor Function After Facial Nerve Injury Requires Intact Sensory Input*

To elucidate the role of trigeminal afferents (infraorbital nerve) which provide the sensory innervation of the vibrissal muscles, we operated animals for FFA, deprived them of trigeminal input (excision of the ipsilateral infraorbital nerve) and subjected them to MS (FFA + ION-ex + MS). Animals subjected to FFA plus excision of the ipsilateral infraorbital nerve without mechanical stimulation served as controls.

MS failed to restore vibrissal function in animals which had facial nerve injury as well as elimination of sensory input (FFA + ION-ex + MS). Moreover, MS led to a worsening of function with a reduction of whisking amplitude in animals that received MS (FFA + ION-ex + MS: 14° ± 5.5°) compared to those that did not (FFA + ION-ex: 22° ± 3.4°; Table 2.6).

This assessment of vibrissal function correlated with the degree of polyinnervation. Thus, rats with FFA + ION-ex and FFA + ION-ex + MS consistently had poor function and a high percentage of polyinnervated end plates (43 ± 9.4 % and 51 ± 10 %, respectively; Table 2.8). By contrast, rats with FFA + MS had normal vibrissal function, and the degree of polyinnervation end plates was significantly smaller (22 ± 5 %).

3.4 Direct Stimulation of the Trigeminal and Facial Nerves by Electric Current to the Vibrissal Muscles Fails to Improve Quality of Target Reinnervation and Does Not Promote Recovery of Vibrissal Function

3.4.1 Electrical Stimulation of the Vibrissal Muscles Does Not Promote Recovery of Whisking

We first examined the effects of electrical stimulation on the recovery of function. Compared to intact animals, vibrissal motion was poor in rats receiving sham stimulation (FFA + SS, group 4) or postoperative ES (FAA + ES; group 5) of the vibrissal muscles. The mean amplitude was reduced to 16 % and 20 % of normal (fourth column in Table 2.9) and the angular velocity to 26 % and 17 %, respectively (fifth column in Table 2.9).

We also noted a specific time course of changes in the intensity of muscle contractions throughout the 2-month period of ES in which the stimulation parameters remained unchanged. Muscle contractions and vibrissal movements were readily visible during the first postoperative week. Thereafter, they declined to zero (probably due to anterograde Wallerian axonal degeneration) and appeared again as reinnervation of end plates took place after 2–3 weeks. After the third postoperative week, it became increasingly difficult to elicit muscle contractions similar to those observed at the beginning of the treatment period (data not shown).

3.4.2 A High Degree of Collateral Axonal Branching Occurs Regardless of ES

Next we used triple retrograde labeling to assess the projection patterns of motor axons from the facial nucleus through its different motor rami as well as the degree of axonal branching (Dohm et al. 2000). In intact animals, motoneurons with axons entering the zygomatic, buccal, or marginal mandibular ramus were localized in the dorsal, lateral, and intermediate facial subnuclei, respectively (Fig. 2.11a, b; Semba and Egger 1986). No double- or triple-labeled motoneurons were observed because intact motoneurons send only one unbranched axon to one of the facialis rami (Fig. 2.11b). Thus, the index of axonal branching in the facial nerve trunk of intact animals, calculated from the zygomatic motoneurons (sum of the percentages in the third and fourth column of Table 2.10), was 0 %.

Two months after facial nerve cut/anastomosis and sham stimulation (FFA + SS, group 4; Fig. 2.11c), myotopic organization into subnuclei was no longer observed, that is, all retrogradely labeled motoneurons were scattered throughout the facial nucleus (Fig. 2.11d). The same phenomenon was observed when FFA was combined with ES (group 5). The lack of myotopy presumably resulted from poor axonal pathfinding and misdirection of the now highly branched regenerating axons after transection and suture of the facial nerve. No fascicular orientation in the zygomatic, buccal, or marginal mandibular branches occurred.

Double and triple labeling was also commonly observed (Fig. 2.11d) and is explained by multiple axonal branches originating from individual perikarya which grow simultaneously into different rami (i.e., zygomaticus, buccalis, and/or marginalis mandibulae); such sprouts retrogradely transported the different fluorescent dyes to their parent motoneurons in the facial nucleus.

In addition, robust collateral branching at the lesion site resulted in the retrograde labeling of more motoneuronal perikarya in each of the individual facial nerve branches than in intact animals (Table 2.10). Such increases in numbers of axons or their branches in turn lead to hyperinnervation of peripheral muscle targets (Angelov et al. 1996; Streppel et al. 1998).

Retrograde tracing did not reveal any changes in the total numbers of single-labeled motoneurons (i.e., DiI-only + FG-only + FB-only; columns 1, 5, and 6; Table 2.10) in the facial nucleus. The total number of single-labeled cells was 2,184 ± 242 in intact rats (group 1), 4,281 ± 830 after sham stimulation (FFA + SS, group 4), and 4,731 ± 756 following ES (FFA + ES, group 5).

Our explanation for the high numbers of single-labeled motoneurons is as follows. In intact animals, axons in individual rami do not branch, and labeling from specific rami (zygomatic, buccal, and marginal mandibular) yields consistent motoneuron numbers with low standard deviations (Table 2.10). The elevated numbers of motoneurons following labeling of the buccal (FG-only) and marginal mandibular (FB-only), but not the zygomatic, branches (Table 2.10) indicate that other axons must have sprouted and entered these rami. Axons from the two other branches of the facial nerve (posterior auricular and cervical), while not labeled in intact animals, must presumably have sprouted into the zygomatic, buccal, and marginal mandibular rami (Table 2.10).

In the present study, the index of axonal branching following FFA and either of the treatments (SS or ES) was 57–62 % (sum of the percentages of DiI + FG and DiI + FB retrogradely labeled perikarya in the third and fourth column of Table 2.10). In summary, the complete lack of myotopic organization and a consistently elevated degree of axonal branching suggested that ES did not influence axonal projection patterns (Fig. 2.11d–g).

3.4.3   ES Does Not Reduce Polyinnervation of the Motor End Plates

In intact animals, all motor end plates were monoinnervated (Fig. 2.2c). After facial nerve injury, the proportion of polyinnervated motor end plates was 49 ± 9.4 % after sham stimulation (FFA + SS, group 4) and 55 ± 14 % after ES (FFA + ES, Fig. 2.2b; Table 2.11). Thus, ES failed to improve the quality of muscle reinnervation.

3.4.4   ES Reduces the Number of Motor End Plates

A very interesting and novel finding is the massive reduction, compared with other treatments, in the number of motor end plates, identified by the alpha-bungarotoxin-binding AChR, in the levator labii superioris muscle at 2 months after ES. The total number of motor end plates observed in animals subjected to SS was 1,398 ± 415. Following ES, this number was reduced to approximately 24 % of the value in the SS group.

This observation has several possible explanations. First, many muscle fibers in the electrically stimulated vibrissal muscles have not been innervated, have degenerated, and have disappeared rapidly (Schmalbruch et al. 1991; Dedkov et al. 2001). Alternatively, it is quite possible that the vast majority of muscle fibers persisted in a strongly atrophied, denervated state after ES. It has been shown in humans that muscle fibers persist in an atrophic state for several years in denervated muscles before they are replaced by connective tissue (Sunderland 1950; Schwarting et al. 1984). The observation that the diameters of most muscle fibers in FFA + ES rats appeared smaller than those in muscles of animals subjected to FFA + SS speaks in favor of this possibility. Finally, the atrophied fibers may represent newly formed fibers that regenerated in the absence of the nerve and were therefore without end plates. Accordingly, ES would have promoted regeneration of muscle fibers.

To find more direct evidence for denervated atrophied muscle fibers, we stained representative sections from all electrically stimulated animals with both alpha-bungarotoxin and anti-AChE (data not shown). We expected to see both innervated end plates (labeled by both alpha-bungarotoxin- and AChE) and denervated end plates which are outlined by AChE immunoreaction only. The AChE can persist, in contrast to the AChRs, for months after denervation of muscles limb and trunk muscles (Lomo and Slater 1980; Decker and Berman 1990). Surprisingly, our analysis showed that postsynaptic AChE was present only in association with postsynaptic nicotinic AChRs. The only explanation for this unexpected result was that the end plates in facial muscles, in contrast to limb musculature (Gordon et al. 2007, 2008), lose both end plate markers, AChE and AChRs, rapidly, within less than 2 months.

To prove this possibility, we permanently denervated all vibrissal muscles in a group of rats ($n = 16$) by resection of the facial nerve. Two months after the operation, the levator labii superiors muscle was indeed completely denervated as indicated by the reduction in the total number of motor end plates and by the absence of beta-tubulin-positive axons.

Altogether, our results show that the ES treatment leads to partial muscle reinnervation after facial–facial anastomosis, a procedure normally followed by complete muscle reinnervation (Angelov et al. 1996).

Chapter 4
Discussion

4.1 Mild Indirect Stimulation of the Trigeminal Afferents by Removal of the Contralateral Vibrissal Hairs Has a Beneficial Effect on Motor Recovery

In this first major set of experiments, we show that, after injury and regeneration of both the facial (motor) and infraorbital (sensory) nerves, noninvasive interventions which ensure a forced use of ipsilateral vibrissae (VS), or which manually stimulate (MS) the whisker pads, improve whisking compared to no treatment; VS followed by MS similarly improves whisking. For all three treatments, functional recovery is associated with reduced polyinnervation of vibrissal muscles. In contrast, regardless of the type of treatment, collateral axonal branching is equally high, whereas synaptic input to facial nucleus motoneurons does not differ from intact animals.

4.1.1 Importance of Sensory Fiber Regeneration for Motor Axonal Regrowth

Following damage to a purely motor nerve, functional recovery is generally better than after damage to mixed (sensory and motor) nerves (Lundborg 2005; Sinis et al. 2005; Kelly et al. 2007). One possibility for the poor recovery is that a lack of sensory input deprives motoneurons from trophic support. Furthermore, motoneuron deafferentation and atrophy can occur (Brännström and Kellerth 1999). Given the extreme disturbances in sensory and motoneuronal excitability following damage to sensory afferents (Devor et al. 1989) as well as excessive sprouting (Shaw and Bray 1977; Diamond et al. 1987) and loss of GABAergic inhibitory control (Castro-Lopes et al. 1993), it is not surprising that recovery in animals is poor when deafferentation is permanent (Pavlov et al. 2008). Interestingly, in the study by Pavlov et al. (2008), MS worsened the functional outcome. The finding raises the possibility that MS

E. Skouras et al., *Stimulation of Trigeminal Afferents Improves Motor Recovery After Facial Nerve Injury*, Advances in Anatomy, Embryology and Cell Biology 213, DOI 10.1007/978-3-642-33311-8_4, © Springer-Verlag Berlin Heidelberg 2013

"overloads" a system that, due to deafferentation, is already hyperexcitable and possibly leads to irretrievable damage (Pavlov et al. 2008).

In the present study, sensory fiber regeneration was facilitated by end-to-end suture of the ipsilateral infraorbital nerve and confirmed by retrograde labeling with Fast Blue (see last section in Results). However, this labeling technique precluded analysis on the accuracy of sensory regrowth. Poor quality sensory reinnervation may have accounted for the reduced functional recovery in the current study compared to when the infraorbital nerve remains intact (Angelov et al. 2007). For example, regenerating sensory axons may form a neuroma which produces sustained long-term ectopic impulses (Devor et al. 1989; Munger and Renehan 1989). Other explanations include a numerical imbalance between regenerated A beta and C fibers (Arvidsson et al. 1986; Woolf et al. 1992; Wilson and Kitchener 1996) and/or loss of afferent inhibitory control due to selective GABAergic cell depletion (Castro-Lopes et al. 1993). The above changes may fail to provide normal afferent input not only to the facial motoneurons but also to the sensory and motor cortex, with an ensuing inability to maintain appropriate levels of excitability within the facial–trigeminal loop (Sosnik et al. 2001; Minnery and Simons 2003). We do not know the extent, if any, to which our different treatments altered the above factors, but regardless, our data suggest that noninvasive stimulation is beneficial when sensory afferents are damaged but have regenerated.

4.1.2 Influence of Synaptic Coverage on Axonal Regrowth and Quality of Target Reinnervation

Following facial nerve injury, synaptic terminals rapidly detach from motoneurons (Graeber and Kreutzberg 1988). However, "synaptic stripping" is reversible if target reinnervation occurs (Neiss et al. 1992; Mader et al. 2004). We are confident that the majority of synapses observed are located on the facial alpha-motoneuron bodies/dendrites because interneurons or gamma-motoneurons are virtually absent in the facial nucleus (Sherwood 2005). Since essentially all motoneurons survive axotomy and reinnervate peripheral muscles (Gordon et al. 2004; Moran and Graeber 2004; Raivich and Makwana 2007), the similar levels of presynaptic input across all groups suggest that such input is unaffected by any of the treatments and that it also does not change if no treatment is given. However, as for any method of assessing synaptic numbers, we were unable to differentiate whether the changes in synaptic input occurred at the soma or at the dendrites and whether they involved specific excitatory, inhibitory, or modulatory synapses.

An interesting finding was that the CLSM produced similar results as the morphological filtering of the WFM images, not only in relative terms, but also in absolute values. The numbers of synapses per motoneuron measured from the WFM images were very slightly lower than those obtained with CLSM because in the former analysis a wider population of perikaryal sizes was considered, while

CLSM focused only on the maximal motoneuron perimeters. These differences could be flattened completely when the linear synaptic densities were considered. The proposed counting on morphologically filtered WFM images is almost completely automated and thus less time consuming and less sensitive to operator-dependent errors. This renders this method a reliable and relatively quick alternative to the CLSM for the quantification of synaptophysin and other immunostained granular material on histological images.

Distinguishing the benefits of potential treatments on mixed nerves is clearly challenging. Furthermore, the precise mechanisms underlying the interventions we tested are unknown. However, ensuring ipsilateral vibrissal use would presumably have involved predominantly sensory input from the skin. By contrast, although proprioceptors are not present in mystacial musculature (Stal et al. 1987; Rice et al. 1997; Whitehead et al. 2005), stroking the whisker pads would provide afferent feedback from the skin as well as stimulation of muscle tissue and circulation thereby limiting loss of muscle mass and minimizing fibrosis (McCulloch and Nelson 1995). Nevertheless, the combined facial and trigeminal nerve injury model has allowed us to distinguish a hierarchy of recovery after facial nerve injury depending on (1) the extent of sensory afferent input and (2) the stimulation of the whisker pads. The best outcome is observed when a facial nerve (purely motor) injury is treated by manual stimulation (Angelov et al. 2007). If, however, as we have shown here, the injury also involves sensory afferents which are allowed to regenerate, some benefit can be derived from the 3 types of stimulation tested. By contrast, animals which do not receive stimulation after facial nerve injury have very little functional recovery (Angelov et al. 2007), while those that have facial nerve injury as well as complete loss of sensory afferent input show the least recovery (Pavlov et al. 2008). However, whatever the level of sensory involvement, when functional improvements occur, they appear to be consistently associated with reduced polyinnervation of target muscles.

4.1.3   Clinical Application

Depicting one more time the importance of regenerating sensory afferents to recovery of motor functions, the present results appear interesting with regard to the feasibility of sensory stimulation for treatment of human patients. Our results show that performance of sensory stimulation (VS) alone or prior to manual stimulation (VS/MS) after the combined (trigeminal plus facial) nerve lesion is not superior for the recovery of motor function when compared to manual stimulation (MS) alone. This partially negative result has one serious advantage for the clinical practice: Physiotherapists do not have to look for or elaborate special methods for sensory stimulation (e.g., moist heat) of paralyzed human mimic muscles. Easing soreness, providing a degree of motion, and increasing blood circulation, the manual stimulation (massage) itself is simultaneously a perfect sensory stimulation of the trigeminal afferents.

The important clinical point here is to know that the method of manual/sensory stimulation is also beneficial in patients with double (facial and trigeminal) nerve lesions. Indeed, using the opposite thumb on the inside of the cheek and the 2 and 3 digits on the facial skin, patients are usually taught to draw the tissues toward the mouth. Usually, most of them report increased comfort and mobility after several weeks of practice (Mosforth and Taverner 1958).

4.2 Beneficial Effect of the Intensive Indirect Stimulation of the Trigeminal Afferents by Excision of the Contralateral Infraorbital Nerve

4.2.1 Removal of the Contralateral Trigeminal (ION) Input Attenuates the Degree of Collateral Axonal Branching Within the Transected Buccal Branch of the Facial Nerve

In this chapter, we selected the vibrissal area as representative of the morpho-functional entity "facial muscles," "trigeminal nerve," "trigeminal nucleus," "facial nucleus," "facial nerve," "facial muscles" and performed surgery on the buccal nerve. Employing this model, we took advantage of:

1. Readily observable verification of postoperative paralysis and recovery by the rhythmical vibrissae movements (Kujawa and Jones 1990)
2. Single and well-established sensory innervation by the infraorbital nerve (Jacquin et al. 1993; Munger and Renehan 1989; Rice et al. 1993)
3. Single and well-established motor nerve supply by the buccal nerve (Dörfl 1985; Klein and Rhoades 1985; Semba and Egger 1986; Hinrichsen and Watson 1984)

A brief review of the results obtained after all three kinds of surgery (BBA only, BBA plus excision of the ipsilateral infraorbital nerve, and BBA plus excision of the contralateral infraorbital nerve), shows the following general conclusions:

1. None of the operation leads to an obvious neuronal loss in the lateral facial subnucleus. The total neuron number remains in the range of 1,500–2,000 (Table 3.1).
2. None of the operation succeeds to restore completely the myotopic organization within the lateral facial subnucleus (cf. Fig. 2.8a, c, e with Fig. 2.8b, d, f).
3. Considering the quantitative aspect of the misdirected reinnervation, however, our data demonstrate that the results after BBA plus excision of the contralateral infraorbital nerve are definitely superior to those obtained after BBA only or BBA plus excision of the ipsilateral infraorbital nerve: The two basic quantitative parameters (numbers of FG-labeled and DiI-labeled motoneurons) estimated after BBA plus excision of the contralateral infraorbital nerve are significantly different and closer to the normal initial values than those obtained after the other two types of surgery (Table 3.1).

A careful analysis of the data listed in Table 3.1 shows that the portion of double-labeled (FG + DiI) motoneurons "underwent" the most dynamic changes in response to various types of surgery: Beginning with a zero in intact rats, their percentage reaches 23 % after BBA only and provides a good explanation for the slow recovery of function in this experimental group. After BBA combined with excision of the ipsilateral infraorbital nerve, the portion of double-labeled motoneurons decreases to 13 %, and after BBA plus excision of the ipsilateral infraorbital nerve, to 8 % of the whole motoneuronal pool.

The importance of axonal branching and its effect on restoration of function became even more evident after an additional experiment. Under narcosis, we transected and sutured the main trunk of the facial nerve in two Wistar rats of the same strain. As there was no recovery of vibrissal whisking till 56 DPO (a period when there were 1,430 ± 36 motoneurons projecting into the whisker pad muscles (Angelov et al. 1996)), we supposed that an unknown portion of these motoneurons could have regrown axonal branches not only within the buccal nerve but also within the adjacent zygomatic and/or marginal mandibular nerves. To corroborate this hypothesis, we reoperated the animals and totally removed the zygomatic and marginal mandibular nerves. Within the following week (5–7 DPO), the vibrissae began rhythmical whisking (own unpublished results). Providing strong evidence for the harmful effect of axonal branching, the results of this additional experiment suggest that the leading pathogenetic factor in the misdirected reinnervation is not the simple misrouting of axons, but the redundant axonal branching enabling a single neuron to project to several targets simultaneously.

Thus, reducing the axonal branching or enhancing the elimination (pruning off) of branches, the combination of facial axotomy plus trigeminal lesion provides a new hard-core evidence for a beneficial effect of the combined surgery on target-selective axonal growth. What are the neurobiological grounds for this entirely new approach to avoid the occurrence or at least to diminish the degree of inevitable postoperative misdirected reinnervation?

Unfortunately, whereas the pathophysiology and the clinical consequences of the post-transectional misdirected reinnervation of muscles are extensively documented, little is known about how the occurrence of this phenomenon could be prevented. So far, there exist two contradictory therapeutical strategies. According to the *first strategy*, the chances of a correct post-transectional axonal routing rise proportionally to the increased number of growth cones from the axons in the proximal stump and to the decreased time required by the growth cones to cross the injured zone, grow down the nerve, and reach the muscle targets (Brown et al. 1981; Fu and Gordon 1995).

The second strategy assumes that despite the established abundance of cues able to guide axons, the growth cones themselves choose their own way by releasing proteases that modify their immediate environment. Consequently, only a perfect synchronization between the degenerative changes in the distal nerve stump and the sprouting from the proximal nerve stump would allow a recovery of the original reinnervation (Dodd and Jessell 1988; Ochi et al. 1994). Our present results fully support this strategy and show that injury-triggered alterations of the trigeminal input to the axotomized facial motoneurons do not merely accelerate the outgrowth rate of growth cones. Nevertheless, they reach the original targets more accurately.

4.2.2 Observations on the Recovering Vibrissal Function

The mystacial vibrissae of the rat perform two types of controlled movements during exploration. The first one is the simultaneous sweep of all vibrissae known as "whisking" or "sniffing" (Welker 1964; Semba et al. 1980) repeated 5–11 times per second (Komisaruk 1970; Carvell et al. 1991). The second aspect of motor behavior is the individual action of single vibrissa in specific palpating fashion upon encountering a novel object (Carvell and Simons 1990). The key movements for both types of motor activity are the protraction and retraction of the vibrissal hairs by the piloerector muscles. All muscles are innervated by the buccal facial nerve (Dörfl 1982; Dörfl 1985).

This is why following any surgery on the buccal facial nerve the vibrissae dropped and acquired caudal orientation. According to earlier reports, the blockade of vibrissae movement requires transection of both the buccal and marginal mandibular nerves (Semba and Egger 1986). In approximately 20 % of the animals from the rat strain used, the marginal mandibular nerve splits into a very thin upper ramus projecting into the vibrissal area and into a rather thick lower ramus projecting toward the digastric muscle. In 80 % of the rats, the marginal mandibular nerve lacked a ramus projecting toward the vibrissal area. This inconsistent division practically forced us to neglect the marginal mandibular nerve supply to the vibrissal muscles. Furthermore, following any type of surgery on the buccal nerve, the vibrissae dropped and remained motionless.

In experimental animal groups 2–4, we evaluated only vibrissal rhythmical whisking. Being totally aware of the fact that this type of motor activity represents a relatively crude and unspecific ballistic movement, we believe that its restoration is at any case superior to the motionless spastic state, which inevitable occurs following surgery. This is why we consider the initiation of rhythmical whisking as an early sign for a more specific recovery of piloerector muscle function.

In animal groups 2a–4a, we failed to observe recovery of "palpating" function. All observations, however, were performed without cinematographic measurements of whisker movements. The equipment with the necessary technique is still under way. Nevertheless, our results demonstrate that following transection and suture of the buccal facial nerve plus excision of the contralateral infraorbital nerve, full restoration of rhythmical whisking occurs at 7–10 DPO, which is 14–18 days earlier than in rats which underwent BBA only (full restoration of function at 21–28 DPO). This finding is supported by electrophysiological measurements and anatomical proofs showing a faster recovery of the whisker pad muscles after suture of the buccal nerve combined with excision of the contralateral infraorbital nerve.

Our behavioral observations are entirely consistent with impacts after various single or combined lesions on the facial nerve in rats observed by other authors (see Huston et al. 1990 for a review) and also with some known benefits of physical therapy in human patients (Baumel 1974; Diels 1994; Taub et al. 1993). Unfortunately, the mechanisms of such improvements are still unknown. Therefore, we can offer only speculative theories which try to explain why a trigeminal lesion reduces facial axonal branching.

The first theory assumes that the observed changes of axonal growth are driven by the abnormal excitability of the spinal trigeminal nucleus as a consequence of the injury to the infraorbital nerve. This excitability is the result of complex changes. First, the proximal stump develops neuroma, which is a source of sustained long-term ectopic impulses (Devor et al. 1989). Second, some 15–20 % of the injured sensory cells and fibers degenerate, whereby the small caliber afferents are preferentially affected (Arvidsson et al. 1986; Wilson and Kitchener 1996). Meanwhile, thicker A delta fibers occupy the "vacant room" of the degenerated A beta and C fibers (Woolf et al. 1992). Third, in the corresponding spinal trigeminal nucleus, inhibitory control decreases due to a selective degeneration of GABAergic cells (Castro-Lopes et al. 1993). All these changes underlie an unnatural, injury-triggered hyperexcitation of the trigeminal nucleus neurons. This abnormal activity could affect further the lateral facial subnucleus via the stronger ipsilateral trigemino-facial projection. Thus, efferents from the spinal trigeminal nucleus release an excess of glutamate, which is known to suppress neurite outgrowth (Owen and Bird 1997). Alternatively, since the crossed trigemino-facial projection is definitely "weaker" than the ipsilateral one (Hinrichsen and Watson 1983), the neurotoxic effect of glutamate is not strong enough to cause sufficient glutamate-mediated damage. Furthermore, there are studies showing that GABA enhances neurite outgrowth (Barbin et al. 1993) and this mechanism may operate under the condition that the crossed trigemino-facial connection is indirect and involves GABAergic interneurons (Li et al. 1997). In summary, this first theory assumes that the slower regrowth and the more abundant branching of axons on the side ipsilateral to the combined (facial and trigeminal) surgery are due to hyperstimulation in the ipsilateral trigeminal nucleus.

The second theory assumes that unilateral excision of the infraorbital nerve deprives the facial motoneurons from trigeminal input and "forces" the rat to use the side with intact sensory innervation. Thus, after BBA only, the rat preferentially uses the intact side even though the sensory innervation is intact on the side of lesion. This deprives the lesioned facial motoneurons from trophic inputs initiated by the powerful ipsilateral sensory afferents or achieved via long-term exercise. The unilateral transection of both the facial and infraorbital nerves creates an even greater emphasis on the normal side at the expense of the lesioned side. Consequently, regeneration proceeds more slowly and with abundant axonal branching. In contrast, if the infraorbital nerve is resected contralateral to the facial nerve transection, then the rat preferentially uses the side with the intact sensory innervation. This, however, is the same side as that with the transected facial axons. Thus, it might be that the mighty ipsilateral afferent input strongly stimulates the axotomized motoneurons to elongate axons, to innervate the piloerector muscles, and thus to provide motor basis for vibrissae rhythmical whisking.

Clinical Implications. The method of combined transection of both the facial and trigeminal nerves finds absolutely no application in human patients. Still, the surprisingly good results in restoration of facial function after excision of the contralateral trigeminal nerve suggest that a temporary blockade of the trigeminal fascicles (e.g., by paraneural injections of various anesthetics) during the post-facial surgery period might yield similar results in humans. The results of such exciting clinical trials are still awaited.

4.2.3   Removal of the Contralateral Trigeminal (ION) Input Improves Quality of Whisker Pad Musculature Reinnervation

The principle finding of this chapter is that following a peripheral lesion to the contralateral trigeminal nerve, approximately 41 % of all axotomized facial motoneurons reinnervated their original target group of muscles. Accordingly, our behavioral observations revealed that the best recovery of vibrissal motor performance occurred after combined facial–trigeminal lesions.

4.2.3.1   Sensory–Motor Integration as a Factor for the Motor Regeneration

From clinical experience after reconstructive surgery on the facial nerve, it is well known that conditioning rehabilitation improves the speed and specificity of reinnervation. This has been supported by the seldom occurrence of synkinesia in patients who undergo conditioning rehabilitation.

Activation of the low-threshold mechanoreceptors in the corresponding region of the face has been frequently used as a conditioning stimulus. Accordingly, increased sensory input may exert a twofold effect on the regenerating facial motoneurons (1) It may speed up axonal elongation and (2) it may decrease aberrant innervation. Our present experiments provide data that fit well with this concept. The rat facial nucleus receives a heavy projection from second-order somatosensory neurons situated throughout the spinal trigeminal complex. This projection is mainly ipsilateral so we expected that the ipsilateral trigeminal nerve would play the most important role for the sensory–motor integration at the level of the facial nucleus. However, our data show significantly better regeneration if the buccal nerve transection and suture were accompanied by a lesion of the contralateral trigeminal nerve.

4.2.3.2   Axonal Branching as a Major Component of the Misdirected Reinnervation

Post-transectional "misdirected" or "aberrant" reinnervation of muscles (Sumner 1990) may result from at least 2 different conditions. First, due to malfunctioning axon guidance, a given group of muscles may receive nonspecific reinnervation by a "foreign" axon that has been misrouted along a "false" endoneural tube through a "wrong" nerve fascicle or ramus (Aldskogius and Thomander 1986; Brushart and Seiler 1987; Esslen 1960; Zhao et al. 1992). Alternately, due to the presence of supernumerary branches from the transected and misguided axons (Al-Majed et al. 2000; Mackinnon et al. 1991; Morris et al. 1972; Shawe 1954), the muscle group can be reinnervated by several axonal branches originating from different motoneurons (Ito and Kudo 1994), a state known as "polyneuronal innervation" (Brown et al. 1981; Rich and Lichtman 1989) or "hyperinnervation" (Angelov et al. 1993). Although thought to be transient (Hennig and Dietrichs 1994), this aberrant innervation may persist for extended periods (Mackinnon et al. 1991; Madison et al. 1999) with deleterious effects on function.

It is extremely difficult to eliminate the first situation in which a "fascicular" or "topographic specificity" is achieved (Evans et al. 1991; Mackinnon et al. 1986). So far, it is technically impossible to guide correctly the growth cones of approximately 1,500 axons and their branches originating from the transected buccal branch of the facial nerve. Accordingly during past several years, we have concentrated on improving the misdirection resulting from the second situation, the axonal branching.

Based on earlier reports that a nimodipine-induced acceleration of axonal regrowth reduces the postoperative hyperinnervation (Angelov et al. 1996, 1997), we tested in Chapter 3 whether alterations in the trigeminal input would reduce branching from the transected facial axons. We found that the altered contralateral trigeminal input not only reduced axonal branching but also improved the conditions for recovery of vibrissae whisking. However, this work considered axonal pathfinding only at the bifurcation of the buccal branch (Angelov et al. 1999). Neglecting the terminal intramuscular sprouting, we studied only the regrowth of nodal axonal branches (Brown and Hopkins 1981; Brushart 1993; Duncan and Baker 1987; Jeng and Cogeshall 1984). Applying intramuscular injections in the course of our present experiments, we explored simultaneously both nodal (Brushart and Mesulam 1980) and intramuscular terminal sprouting of axons in the target (Love and Thompson 1999; Rich and Lichtman 1989; Son et al. 1996; Trachtenberg and Thompson 1996).

4.2.3.3   Nature of the Beneficial Effect of the Accompanying Lesion

The combination of facial nerve axotomy and trigeminal lesion provides solid evidence for a beneficial effect of the combined surgery on muscle reinnervation. What could be the neurobiological grounds for this unusual finding?

The infraorbital nerve (ION) is the only source of sensory innervation for the ipsilateral whisker pad (Chiaia et al. 1988; Jacquin et al. 1986, 1993; Munger and Renehan 1989; Rice et al. 1993; Roades et al. 1987, 1991). Anatomical and electrophysiological studies have shown the existence of direct connections between the ipsilateral (Baumel 1974; Hinrichsen and Watson 1983; Isokawa-Akesson and Komisaruk 1987; Travers and Norgren 1983) and contralateral trigeminal and facial nuclei (Erzurumlu and Killackey 1979; Kimura and Lyon 1972; Moller and Jannetta 1986; Pinganaud et al. 1999). Based on these facts, we put forward two hypotheses to explain why a trigeminal lesion improves the accuracy of facial axonal regrowth.

Regarding *the first hypothesis*, growth cones themselves are to a certain degree capable of selecting a pathway in their immediate environment (Dodd and Jessell 1988; Ochi et al. 1994). The changes of axonal growth in the present experiments are driven by unnatural, injury-triggered alterations in the excitability of the trigeminal nucleus neurons as a consequence of the injury to the infraorbital nerve. A similar phenomenon of axonal growth reorganization has already been described in the spinal cord (Brushart et al. 1981). This abnormal excitability is a result of denervation sensitization of second-order sensory neurons (Mendell 1984; Woolf et al. 1992) in combination with sustained long-term ectopic impulses from the proximal stump neuroma (Devor et al. 1989). Another reason could be the decreased inhibition in the corresponding spinal trigeminal nucleus due to axotomy-induced selective degeneration of GABAergic cells (Castro-Lopes et al. 1993). All these changes might increase the firing rate of second-order trigemino-facial neurons via an excessive glutamate release at the ipsilateral trigemino-facial synapses to suppress neurite outgrowth (Owen and Bird 1997). Alternatively, the "weaker" crossed trigemino-facial projection (Hinrichsen and Watson 1983) might attenuate the neurotoxic effect of glutamate. It has been proposed that these events are accompanied by an enhanced production of brain-derived neurotrophic factor and its high-affinity receptor TrkB by the facial motoneurons (Al-Majed et al. 2000; Fu and Gordon 1997).

Regarding our *second hypothesis*, the improved quality of facial muscle reinnervation is due to a forced overuse of the lesioned facial system caused by behavioral demand. The excision of the contralateral ION "forces" the rat to use the side with the intact sensory innervation. This, however, is the side on which the facial axons were transected. Receiving, in this way, an excessive rehabilitation and use-dependent neuroprotection (but no extreme overreliance) very early after the lesion, the axotomized motoneurons are stimulated strongly to elongate axons, to reinnervate the piloerector muscles, and thus to provide the motor basis for the rhythmical whisking of the vibrissae (Angelov et al. 1999). This increase in the speed of reinnervation in turn reduces the excessive collateral axonal branching and indirectly improves the accuracy of reinnervation.

Either way, both factors – the abnormal sensory input and the overuse of an intact trigeminal nerve – may work in concert to provide the optimal tune-up for the facial motoneurons to regenerate properly to their original target. A clue can be found in the marginal changes that we have seen after BBA + ipsi-ION. Although the increased number of accurately regenerated neurons is not significantly higher than after BBA alone, their number ranges between that seen after BBA and after BBA + contra-ION. This could mean that an ipsilateral trigeminal lesion yields an effect which is similar to that caused by a contralateral lesion but with a much smaller magnitude. Despite the unresolved problems deriving from this experimental paradigm, a theoretical postulate is evolving: That the charge of the sensory input during the process of regeneration of facial motoneurons may play a beneficial role with regard to their post-lesional pathfinding.

4.3   Complete Recovery of Motor Function After Direct Stimulation of the Trigeminal and Facial Nerves by Massage of the Vibrissal Muscles

This chapter provides the first controlled experimental evidence for the efficacy of mechanical muscle stimulation to improve functional recovery after facial nerve injury in the rat. By stroking the whiskers, we stimulated their fine vibrissal muscle slings innervated by the facial nerve and thereby achieved a significant reduction of the proportion of polyinnervated motor end plates and full recovery of vibrissal motor performance.

Restoration of useful function after peripheral nerve injury is a major challenge for reconstructive surgery and rehabilitation medicine (Lundborg 2003). Recent clinical findings have indicated that "facial retraining" using physical rehabilitation can partially improve outcome in a variety of conditions involving facial nerve injury such as acoustic neuroma, Bell's palsy, Ramsay Hunt syndrome, and facial nerve anastomosis (Barbara et al. 2003; Van Swearingen and Brach 2003).

4.3.1   Methodological Considerations

Restoration of vibrissal whisking in rodents is a useful model to study functional recovery after facial nerve injury in humans. However, there should be no confusion with facial hairs in humans. To avoid any misinterpretation, we stress that by stroking the whiskers, we also stimulated their fine vibrissal muscle slings which are innervated by the facial nerve. There is no parallel with human facial hairs since arrector pili muscles are absent from the human facial hairs, eyelashes, and eyebrows, the hairs around nostrils, and the external auditory meatus (Bannister 1995). In addition, arrector pili muscles of hairs elsewhere in the human body are innervated by noradrenergic sympathetic axonal terminals and not by peripheral motor axons. Here, we not only show that manual stimulation can restore whisking function after facial nerve injury in rat but also provide evidence for the mechanisms underlying the recovery.

Considering the method of video-based motion analysis of vibrissae motor performance, one could argue that cutting vibrissae might be to some extent an equivalent to the unloading of the respective levator labii superioris muscle (something like the tendon dissection for the case of the skeletal muscles). Furthermore, cutting all but two vibrissae will certainly result in dramatic increase of the functional load on these remaining two vibrissae, their muscles, and motor and sensory neurons and should definitely be followed by plastic changes in muscle and nerve. In spite of that, we are convinced that the removal of all vibrissal hairs except for two of the C-row on each side of the face has only a negligible effect on the measured motor activity. First, vibrissae removal was applied only at the end of the two-month postoperative survival period and just several minutes before videotaping for analysis. Second, even if these several minutes might have been of some significance, we applied the same procedure for all groups of animals listed in Table 3.2. Finally, sequential observations of videotaped vibrissae movements showed no differences in the motor performance between rats with all vibrissal hairs (before clipping) and rats with only two vibrissae (after clipping).

4.3.2 *Importance of the Stimulation Type*

An important and clinically relevant finding of this investigation is that the success of treatment is determined by the type of stimulation. Recovery under enriched environmental conditions stimulating the use of face muscles, in particular those controlling the vibrissae, was completely ineffective. This is surprising given the known stimulating effects of enriched environments on neuronal plasticity and on adaptive responses (Van Praag et al. 2000). Indeed, clinical success has been thought to rely exclusively on plasticity of cortical and subcortical neuronal networks (Sanes and Donoghue 2000). Our results do not necessarily contradict this view. In our experimental paradigm, sensory innervation of the face remains intact and there should be no alteration of the somatosensory cortical representation. Therefore, cortical plasticity may be of primary importance for restoration of sensory, but not of motor function (Bisler et al. 2002). The surprisingly high efficacy of the mechanical stimulation of the muscles indicates the importance of applying functionally relevant stimulation protocols (Beazley et al. 2003; Dunlop and Steeves 2003).

4.3.3 *Possible Mechanisms of the Beneficial Effects*

4.3.3.1 Mechanical Stimulation Prevents Deafferentation of Regenerated Facial Motoneurons

Following facial nerve injury, synaptic terminals rapidly detach from motoneurons, a phenomenon well known as "synaptic stripping" (Blinzinger and Kreutzberg 1968; Graeber and Kreutzberg 1988). This posttraumatic deafferentation is reversible if target reinnervation occurs (Neiss et al. 1992; Guntinas-Lichius et al. 1994; Mader et al. 2004). Quantitative electron microscopic analysis of regenerated cat gastrocnemius motoneurons has, however, revealed that restoration of synaptic inputs is incomplete in several respects (Brännström and Kellerth 1999). Thus, for example, total synaptic frequency (number per unit membrane length) and total synaptic coverage (percent of membrane length covered by synapses) estimated for motoneuron cell somata and proximal, intermediate, and distal dendritic segments recover to 60–81 % and 28–48 % of normal, respectively. Here we provide for the first time quantitative evidence for deficient recovery of synaptic inputs to regenerated motoneurons in the facial nerve injury paradigm. This claim is warranted in the light of several considerations. First, synaptophysin immunohistochemistry combined with gray value-based densitometry is a well-established approach for estimation of presynaptic terminal densities in different brain regions (see, e.g., Masliah et al. 1990; Svensson and Aldskogius 1993; Calhoun et al. 1996; Spiwoks-Becker et al. 2001; Tiraihi and Rezaie 2004). As applied here, this approach provides estimates that are proportional to the numbers of immunopositive structures rather than to the intensity of immunofluorescence. Second, the facial nucleus contains virtually no interneurons and gamma-motoneurons (Sherwood 2005). Thus, all synaptic terminals in this nucleus make synapses on the dendrites and cell bodies of facial alpha-motoneurons, with the exception of a presumably limited number of axoaxonic synapses. Third, the measurements were performed in the lateral facial subnucleus which contains only motoneurons projecting through the buccal branch of the facial nerve that are all axotomized during FFA. Since essentially all these motoneurons survive axotomy and reinnervate peripheral muscles (Moran and Graeber 2004; Raivich and Makwana 2007), alterations in synaptic inputs cannot be associated with cell survival or success of regeneration. As any other method for estimation of synaptic inputs, our approach has its disadvantages. In particular, this method does not allow assessment of changes in synaptic inputs to different compartments of the motoneuron (cell body and different segments of the dendritic tree). Also, it is impossible to differentiate changes in specific excitatory, inhibitory, and modulatory transmitter systems influencing motoneuron functions in different ways. In view of these considerations, it is important to note that inhibitory and excitatory synaptic inputs may be differentially affected. Brännström and Kellerth (1999) have observed that while the overall restoration of synaptic inputs to spinal motoneurons is deficient, over-compensation occurs for some types of synapses, such as S-type (presumably excitatory) boutons on proximal dendrites. Irrespective of these remarks, we consider the provided evidence for deficient afferent input to regenerated facial motoneurons of substantial importance, not least because, as indicated by the observations made on animals with MS, there is a possible link between level of synaptic input and degree of functional recovery (see below). Further quantitative studies on regeneration-related changes in specific transmitter systems and cell circuitries are required to gain deeper insights into the relationship between structural synaptic plasticity and functional outcome after femoral nerve repair.

Furthermore, our results on synaptic densities suggest that MS prevents the FFA-related loss of facial nucleus afferents. Since both degree of functional recovery and axon terminal densities were superior in treated than in control rats without MS, it appears that in addition to the degree of polyneuronal muscle fiber reinnervation, the level of synaptic input is a second structural parameter which correlates with the degree of functional restoration after FFA. To postulate such a structure–function link is tempting but further studies are required, as indicated above, to verify and understand such correlations. In addition, the explanation of the mechanism underlying the observed effect of MS is not simple. We can speculate that loss and incomplete restoration of facial nucleus innervation without MS is primarily due to loss of input from last-order interneurons in the brain stem, including the principal trigeminal nucleus. With MS, the sensory input that is conveyed directly to these neurons via the trigeminal nerve is enhanced, which in turn leads to augmented loss and/or better restoration of the afferent input to the facial nucleus. The anatomical substrate mediating the effect would then be the vibrissal trigeminal loop, that is, a chain of neurons in the trigeminal ganglion, principle sensory trigeminal nucleus, and subcortical central whisking pattern generator (Kleinfeld et al. 1999; Gao et al. 2001; Kis et al. 2004; Nguyen and Kleinfeld 2005; Leiser and Moxon 2007) interconnected via direct or indirect intrafascicular trigemino-facial brainstem projections. Indeed, there is extensive anatomical, electrophysiological, and clinical evidence for involvement of the trigeminal system in generation of facial muscle responses and blink reflexes (Moller and Sen 1990; Valls-Sole et al. 1992; Zerari-Mailly et al. 2001; Hattox et al. 2002). However, this simple scenario remains questionable regarding the diversity of different projections to the facial nucleus including connections with the neocortex, the other nuclei of cranial nerves, and the reticular formation (Dauvergne et al. 2001; Popratiloff et al. 2001). Not least, we should consider the theoretical possibility that local (intramuscular) effects of MS on muscle fibers and Schwann cells (see below) favor stabilization of synaptic contact after axotomy by altering the retrograde signaling in regenerating motoneurons.

4.3.3.2 Mechanical Stimulation Has No Effect on Collateral Axonal Branching

The mechanisms which limit functional recovery after nerve injury are poorly understood. Our model provides unique opportunities to investigate, in parallel to functional recovery, the substrates of dysfunction. We monitored the influence of the mechanical stimulation protocols on two phenomena. The first one, collateral axonal branching and regrowth to incorrect muscles, was not affected by any of the stimulation protocols. This is surprising in view of the common belief that axonal branching and the subsequent misdirection of the collateral branches are the major factors limiting recovery of the mimic muscles (Ito and Kudo 1994; Dohm et al. 2000; Choi and Raisman 2005). Short-term electrical stimulation of the proximal nerve stump immediately after injury has been shown to improve the speed and accuracy of reinnervation in the rat femoral nerve paradigm (Al-Majed et al. 2000), but the functional impact of this treatment is not known.

In line with our recent observations (Guntinas-Lichius et al. 2005b), the results of the present study question once again the functional significance of collateral axonal branching. In all rats that had been subjected to manual mechanical stimulation, similar to those that underwent only facial nerve cut and suture without mechanical stimulation, retrograde labeling displayed extensive collateral axonal branching (Fig. 2.11d; Table 2.7). As estimated by video-based motion analysis, however, recovery of vibrissal movements was complete in manually stimulated rats and poor to nonexistent in the groups lacking mechanical stimulation.

Lack of whisking recovery after reduced axonal branching has previously been observed by us after facial nerve injury (Tomov et al. 2002). Similarly, use of conduits to reconstruct the sciatic nerve after injury and improve accuracy of reinnervation did not improve locomotor performance in rats (Valero-Cabre and Navarro 2002). The conclusion from the existing experimental evidence is that reduced collateral axonal branching does not lead to better functional restoration. It is currently impossible to explain this unexpected finding. We can only speculate that, as a result of use-dependent plasticity in the CNS, recruitment of aberrantly innervating motoneurons may be reduced or modified, by spinal reflex mechanisms and supraspinal control circuitries, so that the functional disturbances due to this abnormality in reinnervation are minimized. Some studies in humans lend support to this possibility. Cells in a motor nucleus can be "reeducated" to subserve new functional use after muscle–tendon transfer (Illert et al. 1986; Wiedemann et al. 1997). Also, the recruitment of single motor units can be modified via descending control mechanisms as indicated by feedback EMG studies (Guntinas-Lichius 2004).

4.3.3.3   Mechanical Stimulation Improves the Quality of Target Muscle Reinnervation

Next, we evaluated the extent of abnormal muscle reinnervation. The significant reduction in the proportion of polyinnervated muscle fibers achieved by manual stimulation correlated with, and may at least in part, explain functional restoration (Fig. 3.3d–f). In the search for critical factor/s promoting recovery, we recently focused on the polyinnervated end plates of muscle fibers (Guntinas-Lichius et al. 2005b). This phenomenon has always been considered as a factor limiting restoration of function (Schröder 1968; Gorio et al. 1983; Barry and Ribchester 1995; Grimby et al. 1989; Trojan et al. 1991; Tam and Gordon 2003). We examined the levator labii superioris muscle for aberrant reinnervation and observed a qualitatively similar picture in all operated rats which lacked mechanical stimulation. That is, there was an abnormally dense meshwork of intramuscular axonal branches and polyinnervated end plates after facial nerve transection and suture and after transection and suture followed by environmental enrichment (Table 2.8). In contrast, the proportion of polyinnervated end plates was significantly reduced in animals with transection and suture followed by manual stimulation alone or followed by manual stimulation together with environmental enrichment. The effects of mechanical muscle stimulation are readily explainable since previous studies have shown that muscle activity imposed artificially during the phase of synaptic formation and consolidation leads to reduction of the intramuscular sprouting (Brown et al. 1981; Tam et al. 2001; Deschenes et al. 2006).

An earlier study has suggested that intramuscular axonal sprouting in response to muscle paralysis occurs because of short-range diffusible sprouting stimuli generated by the inactive muscle fibers (Brown and Ironton 1977). Accordingly, two subsequent reports have shown that direct stimulation of muscle, but not of nerve, inhibits intramuscular sprouting possibly by interaxonal competition-induced counteracting and/or neutralization of sprouting stimuli arising from the denervated muscle fibers (Brown and Holland 1979; Love et al. 2003). A reduction in perisynaptic Schwann cell processes bridging innervated and denervated end plates might also have occurred (Tam et al. 2001).

Sensory input is presumably another essential, yet unexploited, factor influencing motor recovery. In the nerve lesion paradigm used here, motor axons were lesioned but the circuitry conveying sensory information from the facial skin to the facial motoneurons via the trigeminal nerve was intact. One possibility is that manual stimulation enhanced sensory input and exerted positive effects on the cell circuitry. After complete spinal cord transection, motoneurons distal to the injury undergo atrophy, their dendritic trees shrink and become partially deafferented (Spruston et al. 1995; Segev 1998; Vetter et al. 2001). Similarly, axotomy causes deafferentation of motoneurons and changes in their somata and dendritic trees (Blinzinger and Kreutzberg 1968; Lux and Schubert 1975).

Afferent axons may also exert direct trophic effects on the perikarya and dendrites of motoneurons leading to enhanced production of plasticity-associated molecules such as growth-associated protein 43 (GAP-43), synapsin I, cAMP, and brain-derived neurotrophic factor (BDNF) which stimulate dendrite growth and synaptic remodeling (Al-Majed et al. 2004; Pearse et al. 2004). The notion that stimulation of intact networks improves functions in damaged circuitries is supported by follow-up results in human patients: Stimulation of paralyzed facial muscles and electromyography or neuromuscular reeducation programs are very effective for increasing facial muscle control and function even in cases of long-standing paralysis (Barbara et al. 2003; Cronin and Steenerson 2003; Van Swearingen and Brach 2003).

In conclusion, whereas the exact mechanisms linking mechanical stimulation, polyinnervation, and restoration of the muscle function are still unknown, the facial nerve transection model provides a very good system to address this issue in future experiments. The present report provides clear evidence that manual mechanical stimulation of the denervated muscles can "override" the effects of the robust and consistent (Mackinnon et al. 1991; Reynolds and Woolf 1992), but inappropriate (Esslen 1960), axonal regrowth in the target muscles by reducing the degree of polyinnervation. This effect is apparently sufficient for restoration of motor function. Our findings are pertinent to developing rehabilitation strategies for peripheral nerve injury since they suggest that muscle reinnervation rather than misdirected axonal regrowth should be targeted for therapeutic manipulation.

4.3.4   Adverse Effect of Trigeminal Nerve Ablation on Functional Recovery After FFA

One of our main findings was that the quality of motor reinnervation of whisker pads and whisker function was worse when sensory input is damaged. One possible explanation is that ablating sensory input from the vibrissal whisker pads permanently deprives motoneurons of trophic support and thereby severely limits their recovery. Indeed, in the spinal cord, complete transection results in simultaneous loss of both motor and sensory function. Motoneurons distal to the injury become partially deafferented, undergo atrophy, and their dendritic trees shrink, a condition that persists due to lack of spontaneous regeneration (Spruston et al. 1995; Segev 1998; Vetter et al. 2001). Deafferentation and atrophy of motoneurons also occur after peripheral nerve axotomy but, in contrast to spinal cord injury, peripheral changes are reversed following target reinnervation particularly if sensory damage is avoided (Blinzinger and Kreutzberg 1968; Sumner and Watson 1971; Standler and Bernstein 1982; Brännström et al. 1992a, 1992b; Brännström and Kellerth 1998, 1999; Van den Noven et al. 1993).

4.3.5   The Effect of Manual Stimulation Depends on the Integrity of the Trigeminal Sensory System

Clinically, soft tissue massage following facial nerve damage has been shown to result in improved blood flow, facial symmetry, and smiling (Hovind and Nielsen 1974; Beurskens 1990; Frach et al. 1992; Coulson 2005). The question raised by our current work in rat is as follows: Why does manual stimulation of vibrissal muscles improve recovery after facial nerve injury if the sensory system is intact but make it worse in the absence of normal sensory input?

When the sensory system is intact, MS must provide near normal sensory input to the facial motoneurons, as well as to the sensory and motor cortex, thereby maintaining appropriate levels of excitability within the facial–trigeminal loop (Sosnik et al. 2001; Minnery and Simons 2003). Such relative stability may account for direct beneficial effects on the motoneurons by MS. For example, MS could directly affect the denervated muscle fibers by increasing the circulation, reducing fibrosis, and maintaining membrane properties and therefore responsiveness to action potentials once reinnervation has occurred (Schwarting et al. 1984). Since inactive muscles produce abnormally high levels of growth factors, muscles receiving MS may synthesize fewer growth factors and thus limit inappropriate intramuscular axonal sprouting (Tam et al. 2001; Love et al. 2003).

Another possible substrate is the terminal Schwann cell which, after axotomy, migrates from the perineurium, and enlarges and sprouts "bridges" which reach adjacent innervated motor end plates. Such terminal Schwann cell bridges attract intramuscular sprouts from intact axons toward denervated end plates

(Son et al. 1996). Interestingly, terminal Schwann cell processes precede sprouting from the intact intramuscular axons and are thus able to initiate intramuscular axonal sprouting (Dickens et al. 2003). Manual stimulation may limit the extension of terminal Schwann cell processes and their ability to bridge motor end plates. Indeed, both running and electrical stimulations perturb terminal Schwann cell bridge formation (Tam and Gordon 2003; Love et al. 2003).

Another mechanism responsible for the improved outcome after MS is that the sensory input may facilitate motoneuron regeneration by stimulating plasticity in the brain stem. This phenomenon appears crucial for successful functional recovery of motoneurons in the spinal cord (Sulaiman et al. 2002; Guntinas-Lichius et al. 2005a; Galtrey et al. 2007).

However, as shown here, MS of animals with a damage of both the facial and infraorbital nerves leads to an even worse outcome as compared animals whose sensory system remains intact. It is well known that damage to sensory afferents is followed by extreme disturbances in sensory and motoneuronal excitability (Devor et al. 1989; Schwarz et al. 1983; Spielmann et al. 1983; Bowe et al. 1985), as well as excessive sprouting (Shaw and Bray 1977; Diamond et al. 1987) and loss of GABAergic inhibitory control (Castro-Lopes et al. 1993), and it is thus not surprising that recovery was so poor in animals with both facial and infraorbital nerve injury. We speculate that MS "overloads" a system that has already been rendered susceptible to hyperexcitability, possibly leading to excitotoxicity and thus irretrievable damage.

Our combined facial and trigeminal nerve injury model provides an opportunity to distinguish the response of motor and/or sensory nerves to injury. Our results obtained after damage to both motor and sensory nerves, whether separate or mixed, suggest that the sensory damage is ameliorated with priority. Furthermore, therapeutic strategies aiming at facilitating sensory repair are most likely to be different from those aiming at facilitating motor repair.

4.4 Deleterious Effect of the Direct Stimulation of the Trigeminal and Facial Nerves by Application of Electric Current to the Vibrissal Muscles

Here we show in adult rats that electrical stimulation three times per week for 2 months starting one day after end-to-end suture of the transected facial nerve, a purely motor nerve tract, does not improve motor recovery. Intriguingly, ES reduced the number of innervated motor end plates to 20 % of normal values thus causing partial muscle reinnervation. Furthermore, regardless of whether animals received ES or were sham stimulated, both collateral axonal branching and the proportion of polyinnervated motor end plates were elevated.

Anyway, it is by no means certain that the demonstrated devastating effect on reinnervation would be replicated in human muscles, which may tempt orthopedic surgeons and physiotherapists to make unjustified extrapolation from the rat facial nerve model to the human limb muscles. ES is currently applied to counteract the severe muscle atrophy with interstitial fibrosis which may occur during the very long (6 months or more after brachial plexus surgery) period of reinnervation, that is, maintaining muscle fiber size and structure during the period of nerve regrowth, ES should render the outcome more successful.

4.4.1   Rationale to Use Electrical Stimulation for Treatment of Denervated Muscles

The general neurobiological question whether muscle stimulation during the period of reinnervation would be beneficial is still a major unresolved issue. Earlier findings that electrical stimulation leads to reduced intramuscular axonal sprouting in partially denervated muscles (Brown et al. 1981; Tam et al. 2001; Love et al. 2003) have raised concern that muscle reinnervation might be compromised (Eberstein and Eberstein 1996). Indeed, Hennig (1987) has reported diminished degree of reinnervation, but, at the same time, other experiments have shown either positive (Cole and Gardiner 1984; Einsiedel and Luff 1994; Al-Majed et al. 2000; Mendonca et al. 2003; Gordon et al. 2007) or no effects (Herbison et al. 1973).

Despite a lack of sufficient knowledge from animal experiments, ES of muscles has been widely used in human patients as a rehabilitation treatment over decades for a variety of neural injuries. Not surprisingly, clinical experience is neither consistent nor encouraging and mirrors the animal literature with some positive effects (Cole et al. 1991; Williams 1996; Targan et al. 2000) or no effects (Huizing et al. 1981; Waxman 1984; Gittins et al. 1999; Diels 2005).

Clinically, there are few options for treating either peripheral nerves directly damaged by an injury or their subsequently denervated target muscles. Electrical stimulation (ES) is a possible option although, to date, no form of artificial stimulation has been found to match natural activation for precision or fatigue resistance. One protocol is to stimulate the proximal stump of the damaged nerve acutely, that is, at the time of injury. For example, it has been shown that 1h of ES to the proximal nerve stump of femoral nerve in rats improves the anatomical accuracy of sensory axon regeneration (Brushart et al. 2002, 2005). Furthermore, continuous ES of the proximal stump, ranging from 1h to 14 days after injury, also increases the numbers of regenerating sensory axons, a feature associated with elevated GAP-43 and BDNF expression (Geremia et al. 2007). However, neither study included assessment of functional outcomes that might be related to improved regeneration and targeting. Nevertheless, in a clinical setting, it is possible that such structural improvements might lead to advantages in sensory recovery, possibly even alleviation of neuropathic pain, with the possibility of also facilitating return of motor abilities. Indeed, a recent study assessed return of quadriceps function after 1h of ES applied to the proximal stump immediately after transection (Ahlborn et al. 2007). The brief treatment resulted in approximately the same level of recovery but 6 weeks earlier compared to sham stimulation.

In contrast to humans, in laboratory animals such as rodents, fur prevents the effective use of surface electrodes. Electrodes are therefore implanted close to the target muscle thereby providing direct intramuscular stimulation (Brown et al. 1981; Love et al. 2003). Nevertheless, a great deal of controversy still remains with respect to the use of ES with either some benefit (Farragher et al. 1987; Cole et al. 1991; Williams 1996; Targan et al. 2000; Nicolaidis and Williams 2001; Marqueste et al. 2006) or no effect (Mosforth and Taverner 1958; Huizing et al. 1981; Waxman 1984; Moller and Sen 1990; Kuroki et al. 1994; Ishikawa et al. 1996; Gittins et al. 1999; Marqueste et al. 2002; Diels 2005; Dow et al. 2006).

4.4.2 Effect of Electrical Stimulation on the Quality of Muscle Reinnervation

The success of ES appears to depend on the extent of muscle denervation/reinnervation. Electrical stimulation of partially denervated soleus muscle has the positive effects of inhibiting intramuscular sprouting and diminishing motor end plate polyinnervation (Brown et al. 1981; Love et al. 2003).

By contrast, ES can have adverse effects in partially denervated muscles by stimulating voluntary muscle overuse and suppressing the production of chemical mediators required for reinnervation of denervated muscles (Diels 1994; Tam et al. 2001). In addition, ES of partially denervated muscle reduces the spontaneous electrical activity (fibrillation) of denervated muscle fibers which is thought to be a signal for sprouting of the remaining healthy motor nerve (Cohan and Kater 1986; Brown and Holland 1979).

In the current study, rather than implanting electrodes, we minimized the invasiveness of the ES procedure by using acupuncture needles which were placed at some distance from the motor point, that is, the location of the majority of motor end plates. We delivered ES sufficient to activate axons but not muscle fibers throughout the period of denervation and reinnervation. We did not show any benefit in return of whisking function; rather, we showed the opposite in that ES drastically reduced the degree of muscle fiber reinnervation. To our knowledge, this is the first direct demonstration that ES reduces motor end plate reinnervation although there are a number of studies that indirectly support our observation. In vitro, ES significantly increases neuromuscular synapse elimination compared to that observed in non-stimulated cultures (Nelson et al. 1993). In vivo, ES of partially denervated muscles has been also shown to have positive effects (see above).

By contrast, as we have shown previously (Angelov et al. 2007), MS restores normal whisking function and did so without adversely decreasing the number of motor end plates. The relative benefits of electrical versus mechanical stimulation are largely unknown. However, a study using dantrolene sodium, which interferes with excitation–contraction coupling thereby selectively reducing mechanical rather than electrical activity, showed that loss of polyinnervation was slowed and there was a concomitant reduction in force output (Greensmith et al. 1998).

Taken together, the findings suggest that while electrical stimulation might confer benefit in some situations (e.g., denervated muscles of the extremities with larger motor units), it appears to elicit an adverse effect when applied to denervated small and fine muscles of the face.

Chapter 5
Conclusion

After injury and regeneration of both the facial (motor) and trigeminal (sensory) nerves, noninvasive interventions which ensure a forced use of ipsilateral vibrissae, or which manually stimulate the denervated vibrissal muscles, improve whisking function when compared to no treatment. In contrast, direct electric stimulation of the trigeminal and facial nerves has a deleterious effect on the quality of muscle reinnervation and recovery of function.

E. Skouras et al., *Stimulation of Trigeminal Afferents Improves Motor Recovery After Facial Nerve Injury*, Advances in Anatomy, Embryology and Cell Biology 213, DOI 10.1007/978-3-642-33311-8_5, © Springer-Verlag Berlin Heidelberg 2013

References

Ahlborn P, Schachner M, Irintchev A (2007) One hour electrical stimulation accelerates functional recovery after femoral nerve repair. Exp Neurol 208:137–144

Aldskogius H, Thomander L (1986) Selective reinnervation of somatotopically appropriate muscles after facial nerve transection and regeneration in the neonatal rat. Brain Res 375:126–134

Al-Majed AA, Neumann CM, Brushart TM, Gordon T (2000) Brief electrical stimulation promotes the speed and accuracy of motor axonal regeneration. J Neurosci 20:2602–2608

Al-Majed AA, Tam SL, Gordon T (2004) Electrical stimulation accelerates and enhances expression of regeneration-associated genes in regenerating rat femoral motoneurons. Cell Mol Neurobiol 24:379–402

Angelov DN, Gunkel A, Stennert E, Neiss WF (1993) Recovery of original nerve supply after hypoglossal-facial anastomosis causes permanent motor hyperinnervation of the whisker-pad muscles in the rat. J Comp Neurol 338:214–224

Angelov DN, Neiss WF, Gunkel A, Guntinas-Lichius O, Stennert E (1994) Axotomy induces intranuclear immunolocalization of neuron-specific enolase in facial and hypoglossal neurons of the rat. J Neurocytol 23:218–233

Angelov DN, Gunkel A, Stennert E, Neiss WF (1995) Phagocytic microglia during delayed neuronal loss in the facial nucleus of the rat: time course of the neuronofugal migration of brain macrophages. Glia 13:113–129

Angelov DN, Neiss WF, Streppel M, Andermahr J, Mader K, Stennert E (1996) Nimodipine accelerates axonal sprouting after surgical repair of rat facial nerve. J Neurosci 16:1041–1048

Angelov DN, Neiss WF, Gunkel A, Streppel M, Guntinas-Lichius O, Stennert E (1997) Nimodipine-accelerated hypoglossal sprouting prevents the postoperative hyperinnervation of target muscles after hypoglossal-facial anastomosis in the rat. Restor Neurol Neurosci 11:109–121

Angelov DN, Skouras E, Guntinas-Lichius O, Streppel M, Popratiloff A, Walther M, Klein J, Stennert E, Neiss WF (1999) Contralateral trigeminal nerve lesion reduces polyneuronal muscle innervation after facial nerve repair in rats. Eur J Neurosci 11:1369–1378

Angelov DN, Waibel S, Guntinas-Lichius O, Lenzen M, Neiss WF, Tomov TL, Yoles E, Kipnis J, Shori H, Reuter A, Ludolph A, Schwartz M (2003) Therapeutic vaccine for acute and chronic motor neuron diseases: implications for amyotrophic lateral sclerosis. Proc Natl Acad Sci USA 100:4790–4795

Angelov DN, Ceynowa M, Guntinas-Lichius O, Streppel M, Grosheva M, Kiryakova SI, Skouras E, Maegele M, Irintchev A, Neiss WF, Sinis N, Alvanou A, Dunlop SA (2007) Mechanical stimulation of paralyzed vibrissal muscles following facial nerve injury in adult rat promotes full recovery of whisking. Neurobiol Dis 26:229–242

E. Skouras et al., *Stimulation of Trigeminal Afferents Improves Motor Recovery After Facial Nerve Injury*, Advances in Anatomy, Embryology and Cell Biology 213, DOI 10.1007/978-3-642-33311-8, © Springer-Verlag Berlin Heidelberg 2013

Anonsen CK, Trachy RE, Hibbert J, Cummings CW (1986) Assessment of facial reinnervation by use of chronic electromyographic monitoring. Otolaryngol Head Neck Surg 94:32–36

Arvidsson J (1982) Somatotopic organization of vibrissae afferents in the trigeminal sensory nuclei of the rat studied by transganglionic transport of HRP. J Comp Neurol 211:84–92

Arvidsson J, Ygge J, Grant G (1986) Cell loss in lumbar dorsal root ganglia and transganglionic degeneration after sciatic nerve resection in the rat. Brain Res 373:15–21

Baker RS, Stava MW, Nelson KR, May PJ, Huffman MD, Porter JD (1994) Aberrant reinnervation of facial musculature in a subhuman primate: a correlative analysis of eyelid kinematics, muscle synkinesis, and motoneuron localization. Neurology 44:2165–2173

Banfai P (1976) Die angewandte klinische Anatomie des Nervus Facialis. II. Die Anastomosen. HNO 24:289–294

Bannister LH (1995) The Integumental system. In: Bannister LH, Berry MM, Collins P, Dyson M, Dussek JE, Ferguson MWJ (eds) Gray's anatomy, 38th edn. Churchill Liningstone, New York, p 407

Barbara M, Monini S, Buffoni A, Cordier A, Rochetti F, Harguindey A, Di Stadio A, Cerruto R, Filipo R (2003) Early rehabilitation of facial nerve deficits after acoustic neuroma surgery. Acta Otolaryngol 123:932–935

Barbin G, Pollard H, Gaiarsa J, Ben-Ari Y (1993) Involvement of GABA A receptors in the outgrowth of cultured hippocampal neurons. Neurosci Lett 152:150–154

Barry JA, Ribchester RR (1995) Persistent polyneuronal innervation in partially denervated rat muscle after reinnervation and recovery from prolonged nerve conduction block. J Neurosci 15:6327–6339

Baumel JJ (1974) Trigeminal-facial nerve communications. Their function in facial muscle innervation and reinnervation. Arch Otolaryngol 99:34–44

Beazley LD, Rodger J, Chen P, Tee LBG, Stirling RV, Taylor AL, Dunlop SA (2003) Training on a visual task improves the outcome of optic nerve regeneration. J Neurotrauma 20:1263–1270

Bendella H, Pavlov SP, Grosheva M, Irintchev A, Skouras E, Angelova S, Merkel D, Sinis N, Kaidoglou K, Dunlop SA, Angelov DN (2011) Non-invasive stimulation of the rat whisker pad improves recovery of whisking function after simultaneous lesion of the facial and trigeminal nerves. Exp Brain Res 212:65–79

Bento RF, Miniti A (1993) Anastomosis of the intratemporal facial nerve using fibrin tissue adhesive. Ear Nose Throat J 72:663

Berg RW, Kleinfeld D (2003) Rhythmic whisking by rat: retraction as well as protraction of the vibrissae is under active muscular control. J Neurophysiol 89:104–117

Bermejo R, Zeigler HP (2000) "Real-time" monitoring of vibrissa contacts during rodent whisking. Somatosens Mot Res 17:373–377

Bermejo R, Harvey M, Gao P, Zeigler HP (1996) Conditioned whisking in the rat. Somatosens Mot Res 13:225–233

Bermejo R, Houben D, Zeigler HP (1998) Optoelectronic monitoring of individual whisker movements in rats. J Neurosci Methods 83:89–96

Beurskens CHG (1990) The functional rehabilitation of facial muscles and facial expression. In: Castro D (ed) Facial nerve. Proceedings of the sixth international symposium on the facial nerve. Kugler, Amsterdam, pp 509–511

Bisler S, Schleicher A, Gass P, Stehle JH, Zilles K, Staiger JF (2002) Expression of c-Fos, ICER, Krox-24 and JunB in the whisker-to-barrel pathway of rats: time course of induction upon whisker stimulation by tactile exploration of an enriched environment. J Chem Neuroanat 23:187–198

Blinzinger K, Kreutzberg G (1968) Displacement of synaptic terminals from regenerating motoneurons by microglial cells. Z Zellforsch Mikrosk Anat 85:145–157

Bowe CM, Kocsis JD, Waxman SG (1985) Differences between mammalian ventral and dorsal spinal roots in response to blockade of potassium channels during maturation. Proc R Soc Lond 224:355–366

Brännström T, Kellerth JO (1998) Changes in synaptology of adult cat spinal alpha-motoneurons after axotomy. Exp Brain Res 118:1–13

Brännström T, Kellerth JO (1999) Recovery of synapses in axotomized adult cat spinal motoneurons after reinnervation into muscle. Exp Brain Res 125:19–27

Brännström T, Havton L, Kellerth JO (1992a) Changes in size and dendritic arborization patterns of adult cat spinal alpha-motoneurons following permanent axotomy. J Comp Neurol 318:439–451

Brännström T, Havton L, Kellerth JO (1992b) Restorative effects of reinnervation on the size and dendritic arborization patterns of axotomized cat spinal alpha-motoneurons. J Comp Neurol 318:452–461

Bratzlavsky M, van der Eecken H (1977) Altered synaptic organization in facial nucleus following facial nerve regeneration: an electrophysiological study in man. Ann Neurol 2:71–73

Brown MC, Holland RL (1979) A central role for denervated tissues in causing nerve sprouting. Nature 282:724–726

Brown MC, Hopkins WG (1981) Role of degenerating axon pathways in regeneration of mouse soleus motor axons. J Physiol (Lond) 318:365–373

Brown MC, Ironton R (1977) Motor neurone sprouting induced by prolonged tetrodotoxin block of nerve action potentials. Nature 265:459–461

Brown MC, Goodwin GM, Ironton R (1977) Prevention of motor nerve sprouting in botulinum toxin poisoned mouse soleus muscles by direct stimulation of the muscle [proceedings]. J Physiol 267:42P–43P

Brown MC, Holland RL, Hopkins WG, Keynes RJ (1981) An assessment of the spread of the signal for terminal sprouting within and between muscles. Brain Res 210:145–151

Brushart TM (1993) Motor axons preferentially reinnervate motor pathways. J Neurosci 13:2730–2738

Brushart TM, Mesulam MM (1980) Alteration in connections between muscle and anterior horn motoneurons after peripheral nerve repair. Science 208:603–605

Brushart TM, Seiler WA (1987) Selective reinnervation of distal motor stumps by peripheral motor axons. Exp Neurol 97:289–300

Brushart TM, Henry EW, Mesulam MM (1981) Reorganization of muscle afferent projections accompanies peripheral nerve regeneration. Neuroscience 6:2053–2061

Brushart TM, Gerber J, Kessens P, Chen Y-G, Royall RM (1998) Contributions of pathway and neuron to preferential motor reinnervation. J Neurosci 18:8674–8681

Brushart TM, Hoffman PN, Royall RM, Murinson BB, Witzel C, Gordon T (2002) Electrical stimulation promotes motoneuron regeneration without increasing its speed or conditioning the neuron. J Neurosci 22:6631–6638

Brushart TM, Jari R, Verge V, Rohde C, Gordon T (2005) Electrical stimulation restores the specificity of sensory axon regeneration. Exp Neurol 194:221–229

Calhoun ME, Jucker M, Martin LJ, Thinakaran G, Price DL, Mouton PR (1996) Comparative evaluation of synaptophysin-based methods for quantification of synapses. J Neurocytol 25:821–828

Carvell GE, Simons DJ (1990) Biometric analyses of vibrissal tactile discrimination in the rat. J Neurosci 10:2638–2648

Carvell GE, Simons DJ, Lichtenstein SH, Bryant P (1991) Electromyographic activity of mystacial pad musculature during whisking behavior in the rat. Somatosens Mot Res 8:159–164

Castro-Lopes JM, Tavares I, Coimbra A (1993) GABA decreases in the spinal cord dorsal horn after peripheral neurectomy. Brain Res 620:287–291

Chiaia NL, Allen Z, Carlson E, MacDonald G, Rhoades RW (1988) Neonatal infraorbital nerve transection in rat results in peripheral trigeminal sprouting. J Comp Neurol 274:101–114

Choi D, Raisman G (2005) Disorganization of the facial nucleus after nerve lesioning and regeneration in the rat: effects of transplanting candidate reparative cells to the site of injury. Neurosurgery 56:1093–1100

Clatterbuck RE, Price DL, Koliatsos VE (1994) Further characterization of the effects of BDNF and CNTF on axotomized neonatal and adult mammalian motor neurons. J Comp Neurol 342:45–56

Cohan CS, Kater SB (1986) Suppression of neurite elongation and growth cone motility by electrical activity. Science 232:1638–1640

Cole BG, Gardiner PF (1984) Does electrical stimulation of denervated muscle, continued after reinnervation, influence recovery of contractile function? Exp Neurol 85:52–62

Cole J, Zimmerman S, Gerson S (1991) Nonsurgical neuromuscular rehabilitation of facial muscle paresis. In: Rubin LR (ed) The paralyzed face. Mosby-Year Book, St. Louis, MO, pp 107–112

Coulson SE (2005) Physiotherapy rehabilitation following facial nerve paresis. In: Beurskens CHG, van Gelder RS, Heymans PG, Manni J, Nicolai JPA (eds) The facial palsies. Complementary approaches. Lemma, Utrecht, pp 263–274

Cronin GW, Steenerson RL (2003) The effectiveness of neuromuscular facial retraining combined with electromyography in facial paralysis rehabilitation. Otolaryngol Head Neck Surg 128:534–538

Dauvergne C, Pinganaud G, Buisseret P, Buisseret-Delmas C, Zerari-Mailly F (2001) Reticular premotor neurons projecting to both facial and hypoglossal nuclei receive trigeminal afferents in rats. Neurosci Lett 311:109–112

Decker MM, Berman HA (1990) Denervation-induced alterations of acetylcholinesterase in denrvated and nondenervated muscle. Exp Neurol 109:247–255

Dedkov EI, Kostrominova TY, Borisov AB, Carlson BM (2001) Reparative myogenesis in long-term denervated skeletal muscles of adult rats results in a reduction of the satellite cell population. Anat Rec 263:139–154

Deschenes MR, Tenny KA, Wilson MH (2006) Increased and decreased activity elicits specific morphological adaptations of the neuromuscular junction. Neuroscience 137:1277–1283

Devor M, Keller CH, Deerinck TJ, Ellisman MH (1989) Na$^+$ channel accumulation on axolemma of afferent endings in nerve end neuromas in Apteronotus. Neurosci Lett 102:149–154

Diamond J, Coughlin M, Macintyre L, Holmes M, Visheau B (1987) Evidence that endogenous B nerve growth factor is responsible for the collateral sprouting, but not for regeneration of nociceptive axons in adult rats. Proc Natl Acad Sci USA 84:6596–6600

Dickens P, Hill P, Bennett MR (2003) Schwann cell dynamics with respect to newly formed motor-nerve terminal branches on mature (Bufo marinus) muscle fibers. J Neurocytol 32:381–392

Diels HJ (1994) Treatment of facial paralysis using electromyographic feedback—a case study. Eur Arch Otorhinolaryngol :S129–S132

Diels HJ (2005) Current concepts in non-surgical facial nerve rehabilitation. In: Beurskens CHG, van Gelder RS, Heymans PG, Manni JJ, Nicolai JPA (eds) The facial palsies. Complementary approaches. Lemma Publishers, Utrecht, pp 275–283

Dodd J, Jessell TM (1988) Axon guidance and the patterning of neuronal projections in vertebrates. Science 242:692–699

Dohm S, Streppel M, Guntinas-Lichius O, Pesheva P, Probstmeier R, Neiss WF, Angelov DN (2000) Local application of extracellular matrix proteins fails to reduce the number of axonal branches during regeneration. Restor Neurol Neurosci 16:117–126

Dörfl J (1982) The musculature of the mystacial vibrissae of the white mouse. J Anat 135:147–154

Dörfl J (1985) The innervation of the mystacial region of the white mouse. A topographical study. J Anat 142:173–184

Dow DE, Carlson BM, Hassett CA, Dennis RG, Faulkner JA (2006) Electrical stimulation of denervated muscles of rats maintains mass and force, but no recovery following grafting. Restor Neurol Neurosci 24:41–54

Duncan ID, Baker GJ (1987) Experimental crush of the equine recurrent laryngeal nerve: a study of normal and aberrant reinnervation. Am J Vet Res 48:431–438

Dunlop SA, Steeves JD (2003) Neural activity and facilitated recovery following CNS injury: implications for rehabilitation. In: Bresnahan JC (ed) Topics in spinal cord rehabilitation, vol 8. A Thomas Land, St Louis, MO, pp 92–103

Eberstein A, Eberstein S (1996) Electrical stimulation of denervated muscle: is it worthwhile? Med Sci Sports Exerc 28:1463–1469

Eccles JC, Libet B, Young RR (1958) The behaviour of chromatolyzed motoneurons studied by intracellular recording. J Physiol 143:11–40

Einsiedel LJ, Luff AR (1994) Activity and motor unit size in partially denervated rat medial gastrocnemius. J Appl Physiol 76:2663–2671

Erzurumlu RS, Killackey HP (1979) Efferent connections of the brainstem trigeminal complex with the facial nucleus of the rat. J Comp Neurol 188:75–86

Esslen E (1960) Electromyographic findings on two types of misdirection of regenerating axons. Electroencephalogr Clin Neurophysiol 12:738–741

Evans PJ, Bain JR, Mackinnon SE, Makino AP, Hunter DA (1991) Selective reinnervation: a comparison of recovery following microsuture and conduit nerve repair. Brain Res 559:315–321

Farragher D, Kidd GL, Tallis R (1987) Eutrophic electrical stimulation for Bell's Palsy. Clin Rehabil 1:265–271

Ferguson JH (1978) Hemifacial spasm and the facial nucleus. Ann Neurol 4:97–103

Ferreira MC, Besteiro J, Mand Tuma Junior P (1994) Results of reconstruction of the facial nerve. Microsurgery 15:5–8

Frach JP, Osterbauer PJ, Fuhr AW (1992) Treatment of Bell's palsy by mechanical force, manually assisted chropractic adjusting and high-voltage electrotherapy. J Manipulative Physiol Ther 15:596–598

Friede RL, Bischhausen R (1980) The fine structure of stumps of transected nerve fibers in subserial sections. J Neurol Sci 44:181–203

Fu SY, Gordon T (1995) Contributing factors to poor functional recovery after delayed nerve repair: prolonged axotomy. J Neurosci 15:3876–3885

Fu SY, Gordon T (1997) The cellular and molecular basis of peripheral nerve regeneration. Mol Neurobiol 14:67–116

Galtrey CM, Asher RA, Nothias F, Fawcett JW (2007) Promoting plasticity in the spinal cord with chondroitinase improves functional recovery after peripheral nerve repair. Brain 130:926–939

Gao P, Bermejo R, Zeigler HP (2001) Whisker deafferentation and rodent whisking patterns: behavioral evidence for a central pattern generator. J Neurosci 21:5374–5380

Geremia NM, Gordon T, Brushart TM, Al-Majed AA, Verge VM (2007) Electrical stimulation promotes sensory neuron regeneration and growth-associated gene expression. Exp Neurol 205:347–359

Gittins J, Martin K, Sheldrick J, Reddy A, Thean L (1999) Electrical stimulation as a therapeutic option to improve eyelid function in chronic facial nerve disorders. Invest Ophthalmol Vis Sci 40:547–554

Gordon T, Hegedus J, Tam SL (2004) Adaptive and maladaptive motor axonal sprouting in aging and motoneuron disease. Neurol Res 26:174–185

Gordon T, Brushart TM, Amirjani N, Chan KM (2007) The potential of electrical stimulation to promote functional recovery after peripheral nerve injury–comparisons between rats and humans. Acta Neurochir Suppl 100:3–11

Gordon T, Brushart TM, Chan KM (2008) Augmenting nerve regeneration with electrical stimulation. Neurol Res 30:1012–1020

Gorio A, Carmignoto G, Finesso M, Polato P, Nunzi MG (1983) Muscle reinnervation–II. Sprouting, synapse formation and repression. Neuroscience 8:403–416

Graeber MB, Kreutzberg GW (1988) Delayed astrocyte reaction following facial nerve axotomy. J Neurocytol 17:209–220

Graeber MB, Bise K, Mehraein P (1993) Synaptic stripping in the human facial nucleus. Acta Neuropathol (Berl) 86:179–181

Grant GA, Rostomily RR, Kim DK, Mayberg MR, Farrell D, Avellino A, Duckert LG, Gates GA, Winn HR (2002) Delayed facial palsy after resection of vestibular schwannoma. J Neurosurg 97:93–96

Greene EC (1935) Anatomy of the rat. Transactions of the American Philosophical Society, Philadelphia, New Series, Volume XXVII

Greensmith L, Harding DI, Meyer MP, Vrbová G (1998) Mechanical activity is necessary for the elimination of polyneuronal innervation of developing rat soleus muscles. Brain Res Dev Brain Res 110:131–134

Grimby G, Einarsson G, Hedberg M, Aniansson A (1989) Muscle adaptive changes in post-polio subjects. Scand J Rehabil Med 21:19–26

Gundersen HJ (1986) Stereology of arbitrary particles. A review of unbiased number and size estimators and the presentation of some new ones, in memory of William R. Thompson. J Microsc 14:3–45

Guntinas-Lichius O (2004) The facial nerve in the presence of a head and neck neoplasm: assessment and outcome after surgical management. Curr Opin Otolaryngol Head Neck Surg 12:133–141

Guntinas-Lichius O, Neiss WF (1996) Comparison of empirical and estimated efficiency in neuron counting by the physical disector and in volume measurement by Cavalieri's method. Acta Stereol 15:131–139

Guntinas-Lichius O, Mockenhaupt J, Stennert E, Neiss WF (1993) Simplified nerve cell counting in the rat brainstem with the physical disector using a drawing-microscope. J Microsc 172:177–180

Guntinas-Lichius O, Neiss WF, Gunkel A, Stennert E (1994) Differences in glial, synaptic and motoneuron responses in the facial nucleus of the rat brainstem following facial nerve resection and nerve suture reanastomosis. Eur Arch Otorhinolaryngol 251:410–417

Guntinas-Lichius O, Effenberger K, Angelov DN, Klein J, Streppel M, Stennert E, Neiss WF (2000) Contributing factors to better functional recovery after inevitable delayed nerve repair: enhanced regeneration through the predegenerated distal nerve stump. Exp Neurol 162:98–111

Guntinas-Lichius O, Angelov DN, Tomov TL, Dramiga J, Neiss WF, Wewetzer K (2001) Transplantation of olfactory ensheathing cells stimulates the collateral sprouting from axotomized adult rat facial motoneurons. Exp Neurol 172:70–80

Guntinas-Lichius O, Wewetzer K, Tomov TL, Azzolin N, Kazemi S, Streppel M, Neiss WF, Angelov DN (2002) Transplantation of olfactory mucosa minimizes axonal branching and promotes the recovery of vibrissae motor performance after facial nerve repair in rats. J Neurosci 22:7121–7131

Guntinas-Lichius O, Angelov DN, Morellini F, Lenzen M, Skouras E, Schachner M, Irintchev A (2005a) Opposite impacts of tenascin-C and tenascin-R deficiency in mice on the functional outcome of facial nerve repair. Eur J Neurosci 22:2171–2179

Guntinas-Lichius O, Irintchev A, Streppel M, Lenzen M, Grosheva M, Wewetzer K, Neiss WF, Angelov DN (2005b) Factors limiting motor recovery after facial nerve transection in the rat: combined structural and functional analyses. Eur J Neurosci 21:391–402

Hadlock T, Kowaleski J, Mackinnon S, Heaton JT (2007) A novel method of head fixation for the study of rodent facial function. Exp Neurol 205:279–282

Hadlock T, Kowaleski J, Lo D, Bermejo R, Zeigler HP, Mackinnon S, Heaton JT (2008) Functional assessments of the rodent facial nerve: a synkinesis model. Laryngoscope 118:1744–1749

Harding CV, Unanue ER, Slot JW, Schwartz AL, Geuze HJ (1990) Functional and ultrastructural evidence for intracellular formation of major histocompatibility complex class II-peptide complexes during antigen processing. Proc Natl Acad Sci USA 87:5553–5557

Harsh C, Archibald SJ, Madison RD (1991) Double labeling of saphenous nerve neurone pools: a model for determining the accuracy of axon regeneration at the single neuron level. J Neurosci Methods 39:123–129

Hattox AM, Priest CA, Keller A (2002) Functional circuitry involved in the regulation of whisker movements. J Comp Neurol 442:266–276

Heaton JT, Kowaleski JM, Bermejo R, Zeigler HP, Ahlgren DJ, Hadlock TA (2008) A system for studying facial nerve function in rats through simultaneous bilateral monitoring of eyelid and whisker movements. J Neurosci Methods 171:197–206

Hennig R (1987) Late reinnervation of the rat soleus muscle is differentially suppressed by chronic stimulation and by ectopic innervation. Acta Physiol Scand 130:153–60

Hennig R, Dietrichs E (1994) Transient reinnervation of antagonistic muscles by the same motoneuron. Exp Neurol 130:331–336

Herbison GJ, Teng CS, Gordon EE (1973) Electrical stimulation of reinnervating rat muscle. Arch Phys Med Rehabil 54:156–160

Hinrichsen CF, Watson CD (1983) Brain stem projections to the facial nucleus of the rat. Brain Behav Evol 22:153–163

Hinrichsen CF, Watson CD (1984) The facial nucleus of the rat: representation of facial muscles revealed by retrograde transport of horseradish peroxidase. Anat Rec 209:407–415

Hoffman JR, Greenberg JH, Furuya D, Craik RL, Fanelli P, Breslow S, Sheehan S, Ketschek A, Damkaoutis C, Reivich M, Hand P (2003) Rats recovering from unilateral barrel-cortex ischemia are capable of completing a whisker-dependent task using only their affected whiskers. Brain Res 965:91–99

Hovind H, Nielsen S (1974) Effect of massage on blood flow in skeletal muscle. Scand J Rehabil Med 6:74–77

Huizing EH, Mechelse K, Staal A (1981) Treatment of Bell's Palsy. An analysis of the available studies. Acta Otolaryngol 92:115–121

Huston JP, Steiner H, Weiler HT, Morgan S, Schwarting RKW (1990) The basal ganglia-orofacial system: studies on neurobehavioral plasticity and sensory-motor tuning. Neurosci Biobehav Rev 14:433–446

Ijkema-Paassen J, Meek MF, Gramsbergen A (2002) Reinnervation of muscles after transection of the sciatic nerve in adult rats. Muscle Nerve 25:891–897

Illert M, Trauner M, Weller E, Wiedemann E (1986) Forearm muscles of man can reverse their function after tendon transfers: an electromyographic study. Neurosci Lett 67:129–134

Irintchev A, Draguhn A, Wernig A (1990) Reinnervation and recovery of mouse soleusmuscle after long-term denervation. Neuroscience 39:231–243

Irintchev A, Rollenhagen A, Troncoso E, Kiss JZ, Schachner M (2005) Structural and functional aberrations in the cerebral cortex of tenascin-C deficient mice. Cereb Cortex 15:950–962

Ishikawa M, Namiki J, Takase M, Ohira T, Nakamura A, Toya S (1996) Effect of repetitive stimulation on lateral spread of F-waves in hemifacial spasm. J Neurol Sci 142:99–106

Isokawa-Akesson M, Komisaruk BR (1987) Difference in projections to the lateral and medial facial nucleus: anatomically separate pathways for rhythmical vibrissa movement in rats. Exp Brain Res 65:385–398

Ito M, Kudo M (1994) Reinnervation by axon collaterals from single facial motoneurons to multiple muscle targets following axotomy in the adult guinea pig. Acta Anat (Basel) 151:124–130

Jacquin MF, Renehan WE, Klein BG, Mooney RD, Rhoades RW (1986) Functional consequences of neonatal infraorbital nerve section in rat trigeminal ganglion. J Neurosci 6:3706–3720

Jacquin MF, Zahm DS, Henderson TA, Golden JP, Johnson EM, Renehan WE, Klein BG (1993) Structure-function relationships in rat brainstem subnucleus interpolaris. X. Mechanisms underlying enlarged spared whisker projections after infraorbital nerve injury at birth. J Neurosci 13:2946–2964

Jeng C-B, Cogeshall RE (1984) Regeneration of axons in tributary nerves. Brain Res 310:107–121

Jergovic D, Stal P, Lidman D, Lindvall B, Hildebrand C (2001) Changes in a rat facial muscle after facial nerve injury and repair. Muscle Nerve 24:1202–1212

Kelly EJ, Jacoby C, Terenghi G, Mennen U, Ljungberg C, Wiberg M (2007) End-to-side nerve coaptation: a qualitative and quantitative assessment in the primate. J Plast Reconstr Aesthet Surg 60:1–12

Kern H, Hofer C, Modlin M, Forstner C, Raschka-Hogler D, Mayr W, Stohr H (2002) Denervated muscles in humans: limitations and problems of currently used functional electrical stimulation training protocols. Artif Organs 26:216–218

Kimura J, Lyon LW (1972) Orbicularis oculi reflex in the Wallenberg syndrome: alteration of the late reflex by lesions of the spinal tract and nucleus of the trigeminal nerve. J Neurol Neurosurg Psychiatry 35:228–233

Kimura J, Rodnitzky RL, Okawara SH (1975) Electrophysiologic analysis of aberrant regeneration after facial nerve paralysis. Neurology 25:989–993

Kis Z, Rákos G, Farkas T, Horváth S, Toldi J (2004) Facial nerve injury induces facilitation of responses in both trigeminal and facial nuclei of rat. Neurosci Lett 358:223–225

Klein BG, Rhoades RW (1985) Representation of whisker follicle intrinsic musculature in the facial motor nucleus of the rat. J Comp Neurol 232:55–69

Kleinfeld D, Berg RW, O'Connor SM (1999) Anatomical loops and their electrical dynamics in relation to whisking by rat. Somatosens Mot Res 16:69–88

Komisaruk BR (1970) Synchrony between limbic system theta activity and rhythmical behavior in rats. J Comp Physiol Psychol 70:482–492

Kujawa KA, Jones KJ (1990) Testosterone-induced acceleration of recovery from facial paralysis in male hamsters: temporal requirements of hormone exposure. Physiol Behav 48:765–768

Kuroki A, Moller AR, Saito S (1994) Recordings from the facial motonucleus in rats with signs of hemifacial spasm. Neurol Res 16:389–392

Lagares A, Li HY, Zhou XF, Avendano C (2007) Primary sensory neuron addition in the adult rat trigeminal ganglion: evidence for neural crest glio-neuronal precursor maturation. J Neurosci 27:7939–7953

Leiser SC, Moxon KA (2007) Responses of trigeminal ganglion neurons during natural whisking behaviors in the awake rat. Neuron 55:117–133

Li YQ, Takada M, Kaneko T, Mizuno N (1997) Distribution of GABAergic and glycinergic premotor neurons projecting to the facial and hypoglossal nuclei in the rat. J Comp Neurol 378:283–294

Lieberman AR (1971) The axon reaction: a review of the principle features of the perikaryal response to axon injury. Int Rev Neurobiol 14:49–125

Liu DW, Westerfield M (1990) The formation of terminal fields in the absence of competitive interactions among primary motoneurons in the zebrafish. J Neurosci 10:3947–3959

Lomo T, Slater CR (1980) Control of junctional acetylcholinesterase by neural and muscular influences in the rat. J Physiol 303:191–202

Love FM, Thompson WJ (1999) Glial cells promote muscle reinnervation by responding to activity-dependent postsynaptic signals. J Neurosci 19:10390–10396

Love FM, Son YJ, Thompson WJ (2003) Activity alters muscle reinnervation and terminal sprouting by reducing the number of Schwann cell pathways that grow to link synaptic sites. J Neurobiol 54:566–576

Lundborg G (2003) Richard P. Bunge memorial lecture. Nerve injury and repair—a challenge to the plastic brain. J Peripher Nerv Syst 8:209–226

Lundborg G (2005) Nerve injury and repair. Churchill Livingstone, Edinburgh

Lux HD, Schubert P (1975) Some aspects of the electroanatomy of dendrites. Adv Neurol 12:29–44

Machin R, Perez-Cejuela CG, Bjugn R, Avendano C (2006) Effects of long-term sensory deprivation on asymmetric synapses in the whisker barrel field of the adult rat. Brain Res 1107:104–110

Mackinnon SE, Dellon AL, Lundborg G, Hudson AR, Hunter DA (1986) A study of neurotrophism in a primate model. J Hand Surg Am 11-A:888–894

Mackinnon SE, Dellon AL, Obrien JP (1991) Changes in nerve fiber diametersdistal to a nerve repair in the rat sciatic nerve model. Muscle Nerve 14:1116–1122

Mader K, Andermahr J, Angelov DN, Neiss WF (2004) Dual mode of signalling of the axotomy reaction: retrograde electric stimulation or block of retrograde transport differently mimic the reaction of motoneurons to nerve transection in the rat brainstem. J Neurotrauma 21:956–968

Madison RD, Archibald SJ, Brushart TM (1996) Reinnervation accuracy of the rat femoral nerve by motor and sensory neurons. J Neurosci 16:5698–5703

Madison RD, Archibald SJ, Lacin R, Krarup C (1999) Factors contributing to preferential motor reinnervation in the primate peripheral nervous system. J Neurosci 19:11007–11016

Marqueste T, Decherchi P, Dousset E, Berthelin F, Jammes Y (2002) Effect of muscle stimulation on afferent activities from tibialis anterior muscle after nerve repair by self-anastomosis. Neuroscience 113:257–271

Marqueste T, Decherchi P, Desplanches D, Favier R, Grelot L, Jammes Y (2006) Chronis electrostimulation after nerve repair by self-anastomosis: effects on the size, the mechanical, histochemical and biochemical muscle properties. Acta Neuropathol 111:589–600

Masliah E, Terry RD, Alford M, DeTeresa R (1990) Quantitative immunohistochemistry of synaptophysin in human neocortex: an alternative method to estimate density of presynaptic terminals in paraffin sections. J Histochem Cytochem 38:837–844

McCulloch KL, Nelson CM (1995) Electrical stimulation and electromyographic biofeedback. In: Umphred DA (ed) Neurological rehabilitation. Mosby Year Book, St. Louis, MO, pp 853–856

Mendell LM (1984) Modifiability of spinal synapses. Physiol Rev 64:260–324

Mendonca AC, Barbieri CH, Mazzer N (2003) Directly applied low intensity direct electric current enhances peripheral nerve regeneration in rats. J Neurosci Methods 129:183–190

Minnery BS, Simons DJ (2003) Response properties of whisker-associated trigeminothalamic neurons in rat nucleus principalis. J Neurophysiol 89:40–56

Mitchinson B, Martin CJ, Grant RA, Prescott TJ (2007) Feedback control in active sensing: rat exploratory whisking is modulated by environmental contact. Proc Biol Sci 274:1035–1041

Molander C, Aldskogius H (1992) Directional specificity of regenerating neurons after peripheral nerve crush or transection and epineurial suture: a sequential double-labeling study in the rat. Restor Neurol Neurosci 4:339–344

Moller AR, Jannetta PJ (1986) Blink reflex in patients with hemifacial spasm. Observations during microvascular decompression operations. J Neurol Sci 72:171–182

Moller AR, Sen CN (1990) Recordings from the facial nucleus in the rat. Signs of abnormal facial muscle response. Exp Brain Res 81:18–24

Montserrat L, Benito M (1988) Facial synkinesis and aberrant regeneration of facial nerve. Adv Neurol 49:211–224

Moran LB, Graeber MB (2004) The facial nerve axotomy model. Brain Res Brain Res Rev 44:154–178

Moran CJ, Neely JG (1996) Patterns of facial nerve synkinesis. Laryngoscope 106:1491–1496

Morris J, Hudson AR, Weddell G (1972) Study of degeneration and regeneration in the divided rat sciatic nerve based on electron microscopy. II. The developmen t of the "regenerating unit". Z Zellforsch Mikrosk Anat 124:103–130

Mosforth J, Taverner D (1958) Physiotherapy for Bell's palsy. Br Med J 2:675–677

Munger BL, Renehan WE (1989) Degeneration and regeneration of peripheral nerve in the rat trigeminal system: III. Abnormal sensory reinnervation of rat guard hairs following nerve transection and crush. J Comp Neurol 283:169–176

Naumann T, Härtig W, Frotscher M (2000) Retrograde tracing with Fluoro-Gold: different methods of tracer detection at the ultrastructural level and neurodegenerative changes of back-filled neurons in ling-term studies. J Neurosci Methods 103:11–21

Neefjes JJ, Ploegh HL (1992) Intracellular transport of MHC class II molecules. Immunol Today 13:179–184

Neiss WF, Guntinas-Lichius O, Angelov DN, Gunkel A, Stennert E (1992) The hypoglossal-facial anastomosis as model of neuronal plasticity in the rat. Ann Anat 174:419–433

Nelson PG, Fields RD, Yu C, Liu Y (1993) Synapse elimination from the mouse neuromuscular junction in vitro: a non-Hebbian activity-dependent process. J Neurobiol 24:1517–1530

Nguyen QT, Kleinfeld D (2005) Positive feedback in a brainstem tactile sensorimotor loop. Neuron 45:447–457

Nicolaidis SC, Williams HB (2001) Muscle preservation using an implantable electrical system after nerve injury and repair. Microsurgery 21:241–247

Ochi M, Wakasa M, Ikuta Y, Kwong WH (1994) Nerve regeneration in predegenerated basal lamina graft: the effect of duration of predegeneration on axonal extension. Exp Neurol 128:216–225

Olsson T (1980) Somatopetal transport of horseradish peroxidase following incorporation into axonal sprouts and axons in neuromas after nerve transection. Acta Pathol Microbiol Scand Sect A 88:195–199

Owen AD, Bird MM (1997) Role of glutamate in the regulation of the outgrowth and motility of neurites from mouse spinal cord neurons in culture. J Anat 191:301–307

Pavlov S, Grosheva M, Streppel M, Guntinas-Lichius O, Irintchev A, Skouras E, Angelova S, Kuerten S, Sinis N, Dunlop SA, Angelov DN (2008) Manually stimulated recovery of motor function after facial nerve injury requires intact sensory input. Exp Neurol 211:292–300

Pearse DD, Pereira FC, Marcillo AE, Bates ML, Berrocal YA, Filbin MT, Bunge MB (2004) cAMP and Schwann cells promote axonal growth and functional recovery after spinal cord injury. Nat Med 10:610–616

Pinganaud G, Bernat I, Buisseret P, Buisseret-Delmas C (1999) Trigeminal projections to hypoglossal and facial motor nuclei in the rat. J Comp Neurol 415:91–104

Popratiloff AS, Neiss WF, Skouras E, Streppel M, Guntinas-Lichius O, Angelov DN (2001) Evaluation of muscle re-innervation employing pre- and post-axotomy injections of fluorescent retrograde tracers. Brain Res Bull 54:115–123

Prodanov D, Heeroma J, Marani E (2006) Automatic morphometry of synaptic boutons of cultured cells using granulometric analysis of digital images. J Neurosci Methods 151:168–177

Quist BW, Hartmann MJ (2008) A two-dimensional force sensor in the millinewton range for measuring vibrissal contacts. J Neurosci Methods 172:158–167

Raivich G, Makwana M (2007) The making of successful axonal regeneration: genes, molecules and signal transduction pathways. Brain Res Rev 53:287–311

Raivich G, Jones LL, Kloss CU, Werner A, Neumann H, Kreutzberg GW (1998) Immune surveillance in the injured nervous system: T-lymphocytes invade the axotomized mouse facial motor nucleus and aggregate around sites of neuronal degeneration. J Neurosci 18:5804–5816

Raivich G, Bohatschek M, Kloss CU, Werner A, Jones LL, Kreutzberg GW (1999) Neuroglial activation repertoire in the injured brain: graded response, molecular mechanisms and cues to physiological function. Brain Res Brain Res Rev 30:77–105

Reynolds ML, Woolf CJ (1992) Terminal Schwann cells elaborate extensive processes following denervation of the motor endplate. J Neurocytol 21:50–66

Rice FL, Kinnman E, Aldskogius H, Johansson O, Arvidsson J (1993) The innervation of the mystacial pad of the rat as revealed by PGP 9.5 immunofluorescence. J Comp Neurol 337:66–385

Rice FL, Fundin BT, Arvidsson J, Aldskogius H, Johansson O (1997) Comprehensive immunofluorescence and lectin binding analysis of vibrissal follicle sinus complex innervation in the mystacial pad of the rat. J Comp Neurol 385:149–184

Rich MM, Lichtman JW (1989) In vivo visualization of pre- and postsynaptic changes during synapse elimination in reinnervated mouse muscle. J Neurosci 9:1781–1805

Roades RW, Belford GR, Killackey HP (1987) Receptive field properties of rat VPM neurons before and after selective kainic acid lesions of the trigeminal brainstem complex. J Neurophysiol 57:1577–1600

Roades RW, Enfiejian HL, Chiaia NL, Macdonald GJ, Miller MW, McCann P, Goddard CM (1991) Birthdates of trigeminal ganglion cells contributing axons to the infraorbital nerve and specific vibrissal follicles in the rat. J Comp Neurol 307:163–175

Rossiter JP, Riopelle RJ, Bisby MA (1996) Axotomy-induced apoptotic cell death of neonatal rat facial motoneurons: time course analysis and relation to NADPH-diaphorase activity. Exp Neurol 138:33–44

Sadjadpour K (1975) Postfacial palsy phenomena: faulty nerve regeneration or ephaptic transmission? Brain Res 95:403–406

Sanes JN, Donoghue JP (2000) Plasticity and primary motor cortex. Annu Rev Neurosci 23:393–415

Schmalbruch H, al-Amood WS, Lewis DM (1991) Morphology of long-term denervated rat soleus muscle and the effect of chronic electrical stimulation. J Physiol 441:233–241

Schroder JM (1968) Supernumerary schwann cells during remyelination of regenerated and segmentally demyelinated axons in peripheral nerves. Verh Dtsch Ges Pathol 52:222–228

Schwarting S, Schröder M, Stennert E, Goebel HH (1984) Morphology of denervated human facial muscles. Oto-Rhino-Laryngol 46:248–256

Schwarz JR, Bromm B, Spielmann P, Weytjens JL (1983) Development of Na$^+$ inactivation in motor and sensory myelinated nerve fibres of Rana esculenta. Pflugers Arch 398:126–129

Segev I (1998) Sound grounds for computing dendrites. Nature 393:207–208

Semba K, Egger MD (1986) The facial "motor" nerve of the rat: control of vibrissal movement and examination of motor and sensory components. J Comp Neurol 247:144–158

Semba K, Komisaruk BR (1984) Neural substrates of two different rhythmical vibrissal movements in the rat. Neuroscience 12:761–774

Semba K, Szechtman H, Komisaruk BR (1980) Synchrony among rhythmical facial tremor, neocortical "alpha" waves, and thalamic non-sensory neuronal bursts in intact awake rats. Brain Res 195:281–298

Shaw G, Bray D (1977) Movement and extension of isolated growth cones. Exp Cell Res 104:55–62

Shawe GD (1954) On the number of branches formed by regenerating nerve fibers. Br J Surg 42:474–488

Sherwood CC (2005) Comparative anatomy of the facial motor nucleus in mammals, with an analysis of neuron numbers in primates. Anat Rec 287A:1067–1079

Sinis N, Horn F, Genchev B, Skouras E, Angelova SK, Kaidoglou K, Michael J, Pavlov S, Igelmund P, Schaller HE, Kuerten S, Irintchev A, Dunlop SA, Angelov DN (2009) Electrical stimulation of paralyzed vibrissal muscles reduces endplate reinnervation and does not promote motor recovery after facial nerve repair in rats. Ann Anat 191:356–370

Son YJ, Trachtenberg JT, Thompson WJ (1996) Schwann cells induce and guide sprouting and reinnervation of neuromuscular junctions. Trends Neurosci 19:280–285

Sosnik R, Haidarliu S, Ahissar E (2001) Temporal frequency of whisker movement. I. Representations in brain stem and thalamus. J Neurophysiol 86:339–353

Sparrow JR, Kiernan JA (1979) Uptake and retrograde transport of proteins by regenerating axons. Acta Neuropathol 47:39–47

Spielmann RP, Schwarz JR, Bromm B (1983) Oscillating repolarization in action potentials of frog sensory myelinated fibres. Neurosci Lett 36:49–53

Spiwoks-Becker I, Vollrath L, Seeliger MW, Jaissle G, Eshkind LG, Leube RE (2001) Synaptic vesicle alterations in rod photoreceptors of synaptophysin-deficient mice. Neuroscience 107:127–142

Spruston N, Schiller Y, Stuart G, Sakmann B (1995) Activity-dependent action potential invasion and calcium influx into hippocampal CA1 dendrites. Science 268:297–300

Staiger JF, Bisler S, Schleicher A, Gass P, Stehle JH, Zilles K (2000) Exploration of a novel environment leads to the expression of inducible transcription factors in barrel-related columns. Neuroscience 99:7–16

Stal S, Spira M, Hamilton S (1987) Skin morphology and function. Clin Plast Surg 14:201–208

Standler NA, Bernstein JJ (1982) Degeneration and regeneration of motoneuron dendrites after ventral root crush: computer reconstruction of dendritic fields. Exp Neurol 75:600–615

Streppel M, Angelov DN, Guntinas-Lichius O, Hilgers RD, Rosenblatt JD, Stennert E, Neiss WF (1998) Slow axonal regrowth but extreme hyperinnervation of target muscle after suture of the facial nerve in aged rats. Neurobiol Aging 19:83–88

Sulaiman OA, Midha R, Munro CA, Matsuyama T, Al-Majed A, Gordon T (2002) Chronic Schwann cell denervation and the presence of a sensory nerve reduce motor axonal regeneration. Exp Neurol 176:342–354

Sumner AJ (1990) Aberrant reinnervation. Muscle Nerve 13:801–803

Sumner BE, Watson WE (1971) Retraction and expansion of the dendritic tree of motor neurones of adult rats induced in vivo. Nature 233:273–275

Sunderland S (1950) Capacity of reinnervated muscles to function efficiently after prolonged denervation. Arch Neurol Psychiatry 64:755–771

Svensson M, Aldskogius H (1993) Synaptic density of axotomized hypoglossal motorneurons following pharmacological blockade of the microglial cell proliferation. Exp Neurol 120:123–131

Tam SL, Gordon T (2003) Mechanisms controlling axonal sprouting at the neuromuscular junction. J Neurocytol 32:961–974

Tam SL, Archibald V, Jassar B, Tyreman N, Gordon T (2001) Increased neuromuscular activity reduces sprouting in partially denervated muscles. J Neurosci 21:654–667

Targan RS, Alon G, Kay SL (2000) Effect of log term electrical stimulation on motor recovery and improvement of clinical residuals in patients with unresolved facial nerve palsy. Otolaryngol Head Neck Surg 122:246–252

Taub E, Miller NE, Novack TA, Cook EW 3rd, Fleming WC, Nepomuceno CS, Connell JS, Crago JE (1993) Technique to improve chronic motor deficit after stroke. Arch Phys Med Rehabil 74:347–354

Tetzlaff W, Bisby MA, Kreutzberg GW (1988a) Changes in cytoskeletal proteins in the rat facial nucleus following axotomy. J Neurosci 8:3181–3189

Tetzlaff W, Graeber MB, Bisby MA, Kreutzberg GW (1988b) Increased glial fibrillary acidic protein synthesis in astrocytes during retrograde reaction of the rat facial nucleus. Glia 1:90–95

Thomander L (1984) Reorganization of the facial motor nucleus after peripheral nerve regeneration. An HRP study in the rat. Acta Otolaryngol (Stockh) 97:619–626

Tiraihi T, Rezaie MJ (2004) Synaptic lesions and synaptophysin distribution change in spinal motoneurons at early stages following sciatic nerve transection in neonatal rats. Brain Res Dev Brain Res 148:97–103

Titmus MJ, Faber DS (1990) Axotomy-induced alterations in the electrophysiological characteristics of neurons. Prog Neurobiol 35:1–51

Tomov TL, Guntinas-Lichius O, Grosheva M, Streppel M, Schraermeyer U, Neiss WF, Angelov DN (2002) An example of neural plasticity evoked by putative behavioral demand and early use of vibrissal hairs after facial nerve transection. Exp Neurol 178:207–218

Towal RB, Hartmann MJ (2008) Variability in velocity profiles during free-air whisking behavior of unrestrained rats. J Neurophysiol 100:740–752

Trachtenberg JT, Thompson WJ (1996) Schwann cell apoptosis at developing neuromuscular junctions is regulated by glial growth factor. Nature 379:174–177

Travers JB, Norgren R (1983) Afferent projections to the oral motor nuclei in the rat. J Comp Neurol 220:280–298

Trojan DA, Gendron D, Cashman NR (1991) Electrophysiology and electrodiagnosis of the post-polio motor unit. Orthopedics 14:1353–1361

Umemiya A, Araki I, Kuno M (1993) Electrophysiological properties of axotomized facial motoneurones that are destined to die in neonatal rats. J Physiol 462:661–678

Valero-Cabre A, Navarro X (2002) Changes in crossed spinal reflexes after peripheral nerve injury and repair. J Neurophysiol 87:1763–1771

Valero-Cabre A, Tsironis K, Skouras E, Navarro X, Neiss WF (2004) Peripheral and spinal motor reorganization after nerve injury and repair. J Neurotrauma 21:95–108

Valls-Sole J, Tolosa ES (1989) Blink reflex excitability cycle in hemifacial spasm. Neurology 39:1061–1066

Valls-Sole J, Tolosa ES, Pujol M (1992) Myokymic discharges and enhanced facial nerve reflex responses after recovery from idiopathic facial palsy. Muscle Nerve 15:37–42

Van den Noven S, Wallace N, Muccio D, Turtz A, Pinter MJ (1993) Adult spinal motoneurons remain viable despite prolonged absence of functional synaptic contact with muscle. Exp Neurol 123:147–156

Van Praag H, Kempermann G, Gage FH (2000) Neural consequences of enriched environment. Nat Rev Neurosci 1:191–198

Van Swearingen JM, Brach JS (2003) Changes in facial movement and synkinesis with facial neuromuscular reeducation. Plast Reconstr Surg 111:2370–2375

Vaughan ED, Richardson D (1993) Facial nerve reconstruction following ablative parotid surgery. Br J Oral Maxillofac Surg 31:274–280

Veinante P, Deschenes M (2003) Single-cell study of motor cortex projections to the barrel field in rats. J Comp Neurol 464:98–103

Veronesi C, Maggiolini E, Franchi G (2006) Postnatal development of vibrissae motor output following neonatal infraorbital nerve manipulation. Exp Neurol 200:332–342

Vetter P, Roth A, Hausser M (2001) Propagation of action potentials in dendrites depends on dendritic morphology. J Neurophysiol 85:926–937

Waxman B (1984) Electrotherapy for treatment of facial nerve paralysis (Bell's palsy). Health technology assessment reports. National Center for Health Services Research 3:27

Welker WI (1964) Analysis of sniffing of the albino rat. Behaviour 22:223–244

Whitehead J, Keller-Peck C, Kucera J, Tourtellotte WG (2005) Glial cell-line derived neurotrophic factor-dependent fusimotor neuron survival during development. Mech Dev 122:27–41

Wiedemann E, Eggert C, Illert M, Stock W, Wihelm K (1997) Functional electromyography analysis of radial replacement operation. Orthopade 26:673–683

Williams HB (1996) A clinical pilot study to assess functional return following muscle stimulation after nerve injury and repair in the upper extremity using a completely implantable electrical system. Microsurgery 17:597–605

Wilson P, Kitchener PD (1996) Plasticity of cutaneous primary afferent projections to the spinal dorsal horn. Prog Neurobiol 48:105–129

Wineski LE (1985) Facial morphology and vibrissal movement in the golden hamster. J Morphol 183:199–217

Woolf CJ, Shortland P, Coggeshall RR (1992) Peripheral nerve injury triggers central sprouting of myelinated afferents. Nature 355:75–78

Yu WH, Yu MC (1983) Acceleration of the regeneration of the crushed hypoglossal nerve by testosterone. Exp Neurol 80:349–360

Zerari-Mailly F, Pinganaud G, Dauvergne C, Buisseret P, Buisseret-Delmas C (2001) Trigemino-reticulo-facial and trigemino-reticulo-hypoglossal pathways in the rat. J Comp Neurol 429:80–93

Zhao Q, Dahlin LB, Kanje M, Lundborg G, Lu S-B (1992) Axonal projections and functional recovery following fascicular repair of the rat sciatic nerve with Y-tunneled silicone chambers. Restor Neurol Neurosci 4:13–19

Subject Index

E. Skouras et al., *Stimulation of Trigeminal Afferents Improves Motor Recovery After
Facial Nerve Injury*, Advances in Anatomy, Embryology and Cell Biology 213,
DOI 10.1007/978-3-642-33311-8, © Springer-Verlag Berlin Heidelberg 2013